Adrian Alexander Emmert

The Internal Structure of Periglacial Landforms

Adrian Alexander Emmert

The Internal Structure of Periglacial Landforms

Assessments of Subsurface Variations in Permafrost-related and Frost-related Phenomena by Multi-dimensional Geophysical Investigations

Würzburg University Press

Dissertation, Julius-Maximilians-Universität Würzburg
Fakultät für Geowissenschaften, 2020
Gutachter: Prof. Dr. Christof Kneisel, Prof. Dr. Michael Krautblatter

Impressum

Julius-Maximilians-Universität Würzburg
Würzburg University Press
Universitätsbibliothek Würzburg
Am Hubland
D-97074 Würzburg
www.wup.uni-wuerzburg.de

Print on Demand

Coverdesign: Michael Buchta
Foto: Blockgletscher am Piz Üertsch / Adrian Emmert

ISBN 978-3-95826-138-9 (print)
ISBN 978-3-95826-139-6 (online)
DOI 10.25972/WUP-978-3-95826-139-6
URN urn:nbn:de:bvb:20-opus-202437

Abstract

The internal structure of periglacial landforms contains valuable information on the past and present environmental conditions. To benefit from this archive, however, an enhanced understanding of subsurface variations is crucial. It enables an assessment of the influence of the internal structure on prevailing process regimes and to evaluate the sensitivity of different landform units to environmental changes.

This thesis investigates structural variations in the subsurface of (i) rock glaciers, (ii) solifluction lobes, (iii) palsas/lithalsas and (iv) patterned ground that occur between the different landform types, but also between landform units of the same type. Investigated variables include (i) the spatial distribution of permafrost, (ii) the content of ground ice (iii) the origin of detected ground ice occurrences, (iv) the thickness of the active layer and (v) the topography of the frost table.

Multi-dimensional investigations by the geophysical methods Electrical Resistivity Imaging (ERI) and Ground-Penetrating Radar (GPR) were performed in six study areas (a–f): Four of them are located in high-alpine environments in Switzerland and two of them are located in the subarctic highlands of Iceland. Additionally, surface and subsurface temperature values were continuously recorded at selected study sites. At one study site, pF-values, representing the matric potential (or water potential), were recorded.

From a methodological view, this thesis focuses on the application of quasi-3-D ERI, an approach in which data from multiple two-dimensional data sets is combined to create one three-dimensional data set. This permits e.g., a three-dimensional delimitation of subsurface structures and a spatial investigation of the distribution of ground ice. Besides the analysis of field data, this thesis includes a comparison between inversion models produced with different software products, based on two self-created synthetic data sets.

The detection of resistivity structures and reflection patterns provides valuable insights into the internal structure of the investigated landform units: At the high-alpine study site at (a) Piz Nair, a highly variable ice content indicates a complex development of the investigated rock glacier assembly. The local formation of ground ice is attributed to an embedding of surface patches of snow or ice into the subsurface by rockfall. Results of geoelectric monitoring surveys on selected rock glaciers show the influence of seasonal alterations in the internal structure on subsurface meltwater flow.

At the study site at (b) Piz Üertsch, the presented results indicate the occurrences of isolated ground ice patches in the subsurface of a nearly 500 m long rock glacier. Detected characteristics of the internal structure enable a reconstruction of the development of the rock glacier, in which a temporary override of an adjacent glacier tongue on the pre-existing rock glacier is considered crucial for the current distribution of ground ice. However, the presented results show that buried glacier ice is absent in the subsurface of the rock glacier.

Results from a talus-derived rock glacier near the (c) Las Trais Fluors mountain ridge affirm the existence of a water-permeable permafrost table, which was assumed in previous studies. Furthermore, the presented results show that the investigated rock glacier contains large amounts of rockfall deposits.

A joint interpretation of ERI and GPR results from an investigated scree slope at the mountain (d) Blauberg (Furka Pass) reveals characteristic subsurface structures, which enable a differentiation between solifluction lobes and pebbly rock glaciers.

At the subarctic study site (e) Orravatnsrústir, the presented results show that the internal structure of palsas can be used to deduce their current development stage and to assess their past and future development. Results affirm a long history of palsa development in the area around Lake Orravatn, as assumed by previous studies, but further indicate a recent change in environmental conditions.

The investigated occurrences of patterned ground in the (f) proglacial area of the glacier Hofsjökull are currently not influenced by the detected occurrence of permafrost, according to the presented results. Therefore, a temporary formation of pattered ground is assumed, which is linked to the retreat of the glacier.

This thesis shows discrepancies between the internal structure of some of the investigated landform units and the recent environmental conditions. This indicates a delayed adaption and a low sensitivity of the landform units to environmental changes. Findings indicate that the future development of permafrost will be strongly affected by variations in snowfall. Furthermore, the detection of isolated ground ice occurrences at several study sites contradicts the widely assumed efficiency of balancing heat fluxes to create homogenous subsurface conditions in relatively fine-grained subsurface materials.

Kurzfassung

Der strukturelle Aufbau periglazialer Landformen beinhaltet wertvolle Informationen über vergangene und heutige Umweltbedingungen. Um diese Informationen nutzen zu können, muss jedoch ein vertieftes Verständnis für den Zustand der inneren Struktur und möglicher Variationen entwickelt werden. Dieses Wissen ermöglicht beispielsweise eine Abschätzung des Einflusses der inneren Struktur auf das momentan dominierende Prozess-Regime und eine Beurteilung der Sensitivität gegenüber sich verändernden Umweltbedingungen.

Die vorliegende Arbeit untersucht Unterschiede im Aufbau von (i) Blockgletschern, (ii) Solifluktionsloben, (iii) Palsas/Lithalsas und (iv) Frostmusterböden, die zwischen den Landformtypen, aber auch zwischen einzelnen Einheiten desselben Typs bestehen. Betrachtet werden dabei (i) die räumliche Verbreitung von Permafrost, (ii) der Eisgehalt im Untergrund, (iii) die Entstehung von Untergrundeis, (iv) die Mächtigkeit der Auftauschicht sowie (v) die Formung der Frosttafel.

In sechs Untersuchungsgebieten (a–f), davon vier in Hochgebirgsregionen der Schweiz und zwei im subarktischen Hochland Islands, wurden Untersuchungen mittels mehr-dimensionaler geophysikalischer Verfahren, Widerstandsgeoelektrik (ERI) und Bodenradar (GPR), durchgeführt. Zudem wurden an ausgewählten Standorten kontinuierlich Temperaturwerte der Oberfläche und des Untergrunds aufgezeichnet. An einem Standort wurden ergänzend pF-Werte, die die Saugspannung des Porenwassers angeben, aufgezeichnet. Methodischer Schwerpunkt der vorliegenden Arbeit ist die Anwendung von quasi-3-D ERI, einem Ansatz bei dem Datenpunkte mehrerer zweidimensionaler Datensätze zu einem dreidimensionalen Datensatz vereinigt werden. Dies erlaubt beispielsweise eine dreidimensionale Abgrenzung von Untergrundstrukturen und damit eine räumliche Untersuchung der Verbreitung von Untergrundeis. Ergänzend zur Arbeit mit Felddaten enthält die vorliegende Arbeit einen Vergleich zwischen Inversionsmodellen, die auf Basis von zwei selbst erstellten synthetischen Datensätzen mit unterschiedlichen Softwareprodukten generiert wurden.

Durch die Detektion von Widerstandsstrukturen und Reflektionsmustern lassen sich wertvolle Erkenntnisse über den strukturellen Aufbau der untersuchten Einheiten gewinnen: Im hochalpinen Untersuchungsgebiet am (a) Piz Nair weisen stark schwankende Eisgehalte auf eine komplexe Entwicklungsgeschichte der untersuchten Blockgletschergruppe hin. Die lokale Entstehung von Untergrundeis wird auf Verschüttungen oberflächlicher Schnee- oder Eisfelder durch Steinschlag zurückgeführt. An ausgewählten Blockgletschern wird mittels geoelektrischer Wiederholungsmessungen der saisonale Einfluss der inneren Struktur auf den Schmelzwasserabfluss im Untergrund durch Veränderungen der Permafrosttafel gezeigt.

An einem deutlich größeren Blockgletscher im Untersuchungsgebiet am (b) Piz Üertsch zeigen die Ergebnisse dieser Arbeit isolierte Vorkommen von Untergrundeis. Hier kann anhand der inneren Struktur die Entwicklung des Blockgletschers nachvollzogen werden, wobei insbesondere eine zeitweilige Überdeckung des Blockgletschers durch eine benachbarte Gletscherzunge als ausschlaggebend für die lokale Verteilung von Untergrundeis angesehen wird. Die Ergebnisse zeigen, dass kein Gletschereis in den Blockgletscher eingebettet wurde.

Die vorgestellten Ergebnisse der Untersuchungen an einem Blockgletscher nahe des Bergkamms (c) Las Trais Fluors bestätigen die dort in vorherigen Studien angenommene Wasserdurchlässigkeit der Frosttafel. Zudem zeigt der Aufbau des Blockgletschers das Auftreten großer Mengen von Steinschlagablagerungen.

Am untersuchten Schutthang am (d) Blauberg (Furkapass) können durch eine gemeinsame Auswertung der Ergebnisse von ERI und GPR charakteristische Strukturen detektiert werden, durch die sich die dort auftretenden Lobenstrukturen in Solifluktionsloben und Feinmaterial-Blockgletscher (Pebbly Rock Glaciers) unterscheiden lassen.

Im subarktischen Untersuchungsgebiet (e) Orravatnsrústir zeigen die Ergebnisse, dass vom strukturellen Aufbau von Palsas auf deren gegenwärtiges Entwicklungsstadium geschlossen werden kann und dass Rückschlüsse auf vergangene und zukünftige Entwicklungen möglich sind. Die vorgestellten Ergebnisse bestätigen die in vorherigen Studien getroffene Annahme einer lange zurückreichenden Entwicklungsgeschichte der Palsas, weisen aber auch auf sich seit kurzem verändernde Umweltbedingungen hin.

Die untersuchten Frostmusterböden im (f) Gletschervorfeld des Hofsjökull-Gletschers zeigen gegenwärtig keine Beeinflussung durch Permafrost, obwohl ein rezentes Vorkommen von Permafrost angenommen wird. Daher wird eine temporäre Bildung der untersuchten Oberflächenstrukturen angenommen, die an den Rückzug des Gletschers gebunden ist.

Die vorliegende Arbeit zeigt, dass die innere Struktur einiger der untersuchten Landform-Einheiten Diskrepanzen zu den momentanen Umweltbedingungen aufweist. Dies deutet auf eine geringe Sensitivität, beziehungsweise eine verzögerte Anpassung der Landschaftsformung auf sich verändernde Umweltbedingungen hin. Des Weiteren zeigt die vorliegende Arbeit, dass besonders Veränderungen im Schneedeckenauf- und abbau wesentlich zur zukünftigen Entwicklung von Permafrost in den untersuchten Gebieten beitragen werden. Die Beobachtung isolierter Vorkommen von Untergrundeis in mehreren Untersuchungsgebieten steht in Kontrast zur verbreiteten Annahme, dass die ausgleichende Wirkung von Wärmeströmen im Untergrund in feinkörnigem Material besonders stark ist.

Acknowledgments

This work would not have been possible without the great support from many people to which I am much obliged.

I wish to show my gratitude to Prof. Christof Kneisel from the chair of Physical Geography at the University of Würzburg, supervisor of this thesis and my academic mentor, for the opportunity to write this thesis in his research project KN542/13-1, which was funded by the German Research Foundation (DFG). His support did not only cover scientific supervision but also practical help during fieldwork and lifts in his 4WD.

I am further indebted to the staff at the Institute of Geography and Geology at the University of Würzburg for my academic education during bachelor's and master's studies. Thanks to the team at the Chair of Physical Geography of Prof. Roland Baumhauer for the pleasant and productive working environment during my engagement there between 2014 and 2017 and for the support with technical and wheeled equipment. Thanks for the opportunity to purchase both a GPR and an RTK-GNSS system, which strongly enhanced my methodological capabilities. Thanks to my former colleagues for not vetoing against these purchases and sorry for the nearly permanent occupation of the devices.

Geophysical surveying in periglacial environments is a challenging task that can't be overestimated. It requires the interaction of numerous helping hands during fieldwork and the transport of heavy equipment on shoulders and backs for kilometers, in horizontal and vertical direction and often under unpleasant weather conditions. Therefore, I would like to recognize the assistance of Felix Camenzuli, Clemens Emmert, Karl Heinz Emmert, Danilo Fries, Jasmin Gebhard, Jennifer Klemm, Julius Kunz, Sandra Mohr, Alexander Nestle, Jakob Rieser, Armin Rist, Carina Selbach and Julian Trappe.

I would like to thank the municipalities of Bergün/Bravuogn and Celerina/Schlarigna for the permission to use their dirt roads and the AG Luftseilbahn Corviglia–Piz Nair for providing occasional transfer. Thanks to Erika Hiltbrunner and the staff at the Alpine Research Station Furka (ALPFOR) for providing accommodation in a friendly atmosphere and thanks to Thorsteinn Saemundsson for his support during fieldwork in Iceland.

I would like to pay my special regards to Prof. Michael Krautblatter and his team at the Professorship of Landslide Research at the Technical University of Munich (TUM) for the friendly welcoming in 2017 and for guiding me on my first steps at TUM. Furthermore, I wish to thank Prof. Uwe Stilla and Ludwig Hoegner at the Professorship of Photogrammetry and Remote Sensing for giving me the opportunity to continue research in high-alpine environments and my colleagues at FPF, LMF and SiPEO for their support when I was finishing this thesis.

I am thankful to my parents, family and friends who supported me during all the years of my studies.

Contents

List of Figures

List of Tables

Abbreviations

Abbreviation	Description
APIM	Alpine Permafrost Index Map
BAFU	Bundesamt für Umwelt
BERT	Boundless Electrical Resistivity Tomography
BTS	Bottom Temperature
CMP	Common Mid-Point
DEM	Digital Elevation Model
DipDip	Dipole-Dipole
EM	Electromagnetic
ERI	Electrical Resistivity Imaging
ERTM	Electrical Resistivity Tomography Monitoring
GNSS	Global Navigation Satellite System
GPR	Ground-Penetrating Radar
GST	Ground Surface Temperature
LIA	Little Ice Age
MAAT	Mean Annual Air Temperature
MAGST	Mean Annual Ground Surface Temperature
RTK	Real-Time Kinematic
SC	Snow Cover
SRT	Seismic Refraction Tomography
swisstopo	Bundesamt für Landestopografie (CH)
TS	Time Step
TWT	Two-way Traveltime
UAV	Unmanned Aerial Vehicle
WenSl	Wenner-Schlumberger
NAR	Rock Glacier Assembly Piz Nair
UER	Rock Glaciers Piz Üertsch and Piz Blaisun
LTF	Las Trais Fluors Rock Glacier
FUR	Blauberg/Furka Pass
OVR	Orravatnsrústir Palsa Site
HPA	Hofsjökull Proglacial Area

Upper Section: General Abbreviations // Lower Section: Study Sites

Symbols

Symbol	Description
ρ	Specific Resistivity
ρ_a	Apparent Resistivity
Φ	Porosity
m	Cementation Exponent
a	Empirical Constant
V	Electric Potential
I	Intensity of Electric Current
k	Geometry Factor
Φ_d	Data–Model Misfit
Φ_m	Model Roughness
Φ_{total}	Total Data Misfit
λ	Regularization Parameter
d	Measured Data
d_{pred}	Predicted Data
W_d	Weighting Matrix
J	Jacobian Matrix
R	Roughness Matrix
m	Model Parameter
v	EM Velocity
c	Speed of Light
ϵ	Dielectric Constant
R	Reflection Coefficient
A	Signal Attenuation
z	Travel Distance
α	Attenuation Constant
σ	Electrical Conductivity
μ	Magnetic Permeability
r	Radius
h	Depth
λ	Wavelength

Upper Section: ERI // Central Section: Inversion of Resistivity Data // Lower Section: GPR

Chapter 1
Introduction

The first chapter of this thesis outlines its general setting in the context of periglacial geomorphology and its position in the state of current research. At the beginning, definitions are introduced and an overview of periglacial processes and landforms is presented (1.1). This is followed by a description of environmental changes in periglacial areas, both observed and projected, that are associated with climate change (1.2). After the second section, the current state of research on the internal structure of different types of periglacial landforms is summarized (1.3), with a focus on investigations with geophysical methods. The first chapter closes with the objectives of this thesis, explicitly stated as research questions (1.4).

1.1 The Periglacial Domain

The term *periglacial* describes

> *"conditions, processes and landforms associated with cold, non-glacial environments"*
>
> — Van Everdingen (2005).

Corresponding conditions are present in arctic and subarctic areas of high latitude and in mountainous areas of high altitude, which are both characterized by cold temperatures and intense freezing. To delimit the extent of the periglacial zone, Williams (1961) suggested to use a value of 3 °C for the mean annual air temperature (MAAT) as a lower boundary to delimit the area in which periglacial phenomena can be observed. A refined concept subdivides this area by using the −2 °C MAAT isotherm into one section with lower temperatures, in which frost-action conditions dominate, and into another section with higher temperatures, in which frost-action conditions occur, but not in a dominant way (French, 2018). Beside the obligatory occurrence of intense seasonal frost, the occurrence of permafrost is not mandatory to define periglacial environments.

Permafrost is defined as

> *"ground (soil or rock and included ice and organic material) that remains at or below* 0 °C *for at least two consecutive years"*
>
> — Van Everdingen (2005).

Permafrost is present in areas where an intense freezing in winter is stronger than thawing in summer, so that only the uppermost part of the subsurface, the so-called *active layer*, seasonally thaws (Dobiński, 2011) (see fig. 1.1). An approximate outline for the global extent

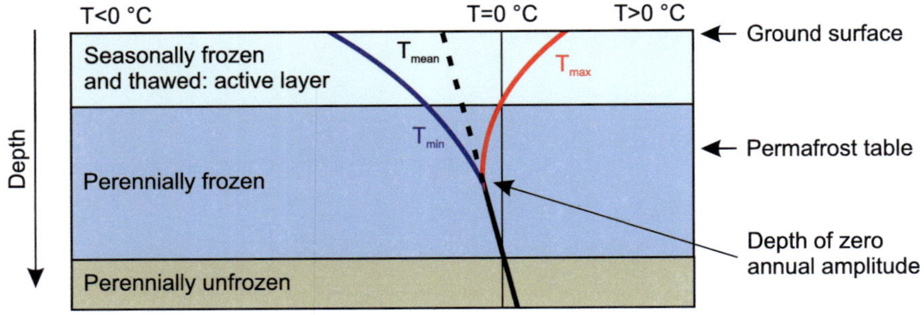

Figure 1.1: Schematic Ground Thermal Regime in Permafrost Areas. Modified Sketch from French (2018).

of permafrost is the position of the −1 °C MAAT isotherm. Following the calculations of Zhang et al. (2008), approximately 23.9 % of the land surface of the northern hemisphere is currently affected by permafrost. The presented definition of permafrost considers the two variables (i) temperature and (ii) time only, but not the occurrence of water. Permafrost can thus be in a dry state or contain small amounts of liquid water (Ballantyne and Murton, 2017). For research in a geomorphologic context, however, the increasing volume of water during phase transition from a liquid to a solid state is one of the key factors that determine frost-related processes and periglacial landform development.

Periglacial processes are geomorphologic processes that are related to intense freezing. They are not exclusively bound to the occurrence of permafrost conditions and most of these processes are not restricted to periglacial environments. Washburn (1979) names (i) process combinations and (ii) process strength as two key variables to characterize periglacial conditions. This view avoids a normative listing of explicitly periglacial processes. However, processes that are commonly associated with periglacial conditions are e.g., frost heave (i.e., surface uplift due to the formation of ground ice), or ice segregation (i.e., formation of ice layers in freezing soils from migrating pore water) (Van Everdingen, 2005).

Periglacial landforms are the visible expressions of such processes. The detection of specific landform types allows conclusions on environmental conditions even in remote areas, like planet Mars, where structures on the planetary surface resemble structures that occur on the Earth's surface (Mellon et al., 2008; Squyres and Carr, 1986). Although there is still no commonly accepted definition of the term *landform*, it represents a key concept in geomorphology (Barsch, 1993; Rhoads and Thorn, 1996). This is because the internal structure of landforms can act as an archive for paleoclimatic and paleoecologic conditions, while knowledge on the current environmental conditions can be used for a classification of landform units. Similar to periglacial processes, an explicit listing of periglacial landforms is neither possible nor useful in a continuous environment (Berthling and Etzelmüller, 2011).

1.2 Climate Change in Periglacial Areas

The aforementioned definition of the term *periglacial* not only includes the thermal conditions explicitly named, but also further climate-related parameters like precipitation. This is because snow can e.g., insulate the subsurface from solar radiation and thereby determine

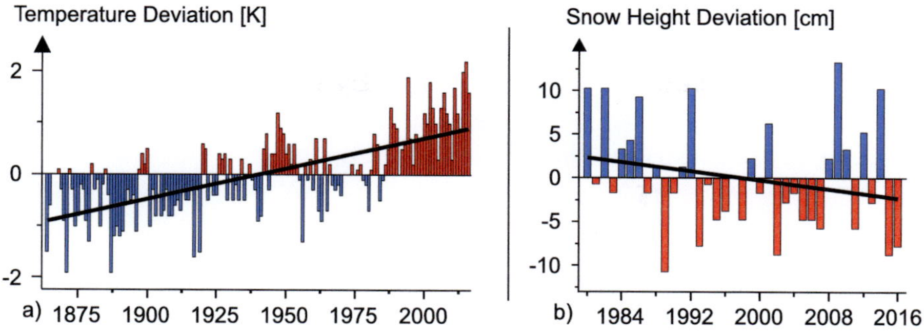

Figure 1.2: Long-Term Development of Environmental Factors in Samedan (CH). a) Trend of MAAT, b) Trend of Snow Height. Data from MeteoSchweiz (2018).

the local energy input into the ground. Hence, the projected environmental changes, such as changes in snowfall or air temperature (see Kovats et al., 2014; Larsen et al., 2014 for details on climate change) will significantly alter the periglacial domain in various ways. These alterations will also modify the internal structure of periglacial landforms, as surface and subsurface conditions are closely linked. Thus, it is important to summarize the expected impacts of the observed and projected environmental changes.

The generally observed increase in surface temperatures is particularly strong in the European Alps. In the *Greater Alpine Region*, defined as the area between 4° E and 19° E as well as between 43° N and 49° N, the increase between the late 19th century and the early 21st century is two times as high as the average increase in the northern hemisphere (Auer et al., 2007). Warming temperatures in the Engadin, where some study sites of this thesis are located (see fig. 2.1), are exemplary illustrated in figure 1.2a. Projections based on the rather balanced A1B emission scenario (Nakicenovic et al., 2000) show that the temperature increase in the Greater Alpine Region will accelerate from 0.25 °C 10 a^{-1} in the first half of the 21st century to 0.36 °C 10 a^{-1} in the second half (Heinrich et al., 2013).

Beside the observed and projected increase in temperature, environmental changes concern snowfall intensity and its temporal pattern. Data from Switzerland shows that the ratio between the number of days with snowfall and the number of days with precipitation decreased between 1961 and 2008 (Serquet et al., 2011). This decrease is particularly strong at lower elevations, but in the periglacial zone, which is assumed to cover elevations between 2000 m a.s.l. and 2700 m a.s.l., the average decrease is still 7.29 %. Regarding data from 1931 to 1999, Laternser and Schneebeli (2003) show a decrease in snow height for many locations in the Swiss Alps after the early 1980s and a simultaneous decrease in snow cover durability. The aforementioned study further shows that an earlier onset of snowmelt in spring has a stronger impact on the shortening of snow cover duration than a later onset of snowfall in autumn. Using again the example of the Engadin, the decrease in snow height is illustrated in 1.2b. Projections on snowfall and snow cover characteristics are more complex than projections on temperature, as interactions between the two variables temperature and precipitation must be considered. Assuming the A2 emission scenario, which corresponds to economic growth (Nakicenovic et al., 2000), Beniston et al. (2011) show that snow-unfavorable conditions will occur more frequently between 2071 and 2100.

Environmental studies from areas of high latitude show similar results as studies from areas of high altitude. The temperature increase in the Arctic, defined here as all areas north of 60° N, was 0.36 °C 10 a^{-1} between 1977 and 2001 (Jones and Moberg, 2003). Long-term observations from Iceland, where two study sites of this thesis are located, show that the magnitude of the temperature increase in Iceland has been comparable to the magnitude of the temperature increase in other parts of northern Europe during most of the 20th century. An exception to this correspondence exists only between the 1940s and the 1980s, when temperature records from Iceland show a distinct cooling (Hanna et al., 2004). Assuming again the A2 emission scenario, climate simulations show that both temperature and precipitation will increase in northern Europe between 2071 and 2100, and that this increase will be particularly strong in winter (Räisänen et al., 2004).

Large-scale projections show a strong decrease in the extent of the permafrost-affected area. Based on the A2 emission scenario, the permafrost-affected area in the northern hemisphere will be reduced by around 90 % between recent times and the year 2100 (Lawrence and Slater, 2005). However, based on the comparable RCP8.5 emission scenario (Riahi et al., 2011), the application of a more sophisticated climate model than the one that was used in the aforementioned study, reduces the value of decrease to 72 % (Lawrence et al., 2012). Regional differences in the projected decrease in permafrost extent exist between high-latitude permafrost areas in the Tibetan Plateau where the decrease is higher, and high-altitude permafrost areas where the decrease is lower (Guo and Wang, 2016). In the Swiss Alps, changes of temperatures and snow cover duration will presumably have a strong impact on the future extent of permafrost, while changes of snow height are assumed to be of minor importance, as a modeling approach of Marmy et al. (2013) shows.

As aforementioned, climate change will alter the local process dynamics in periglacial areas and subsequently affect landform development. To assess the strength of these alterations, knowledge on the sensitivity of the landforms to environmental changes is crucial. This sensitivity is determined by characteristics of the internal structure, like debris composition, the occurrence of ground ice, thickness of the active layer or frost table topography. These parameters interact with each other and can be heterogeneously distributed even over small areas in the subsurface (Lambiel and Pieracci, 2008; Luetschg et al., 2004).

Fine-grained debris is assumed to be more sensitive to climatic changes than coarse-grained debris (Schneider et al., 2012). This is because higher surface temperature values were measured at sites where fine-grained materials are exposed (Hoelzle et al., 1999; Rödder and Kneisel, 2012a), indicating a higher sensitivity (Kääb et al., 2007; Lambiel and Delaloye, 2004). Fine-grained materials further often lack ice supersaturation (Ikeda and Matsuoka, 2006) and contain a higher amount of liquid water, even at subzero temperatures. This is due to a relatively high water retention capacity and the ability of fine-grained materials to reduce the speed of percolating meltwater (Ikeda et al., 2008; Schneider et al., 2013).

The incorporation of remnant ice patches from former glaciations into the internal structure of periglacial landforms can lead to a local increase of ice content and, in case of exposure and subsequent melting, can trigger thermokarst processes (Monnier et al., 2013; Stenni et al., 2007). Despite their sensitivity to thermal disturbances in case of exposure (Haeberli and Vonder Mühll, 1996), remnant ice bodies can persist in case of undisturbed conditions and thereby indicate a rather slow response to climatic changes (Moorman and Michel, 2000; Ribolini et al., 2010). The incorporation of surface accumulations of snow or ice into periglacial landforms, e.g., by rockfall or by debris-rich avalanches, is frequently reported from mountainous areas (Isaksen et al., 2000; Kenner et al., 2017). It can lead to a local increase in ice content and lower debris temperatures towards the freezing point (Haeberli et al., 2006). An enhanced supply with meltwater from the embedded snow and a subsequent refreezing of the meltwater can further result in the development of distinct ice bodies, which are assumed to initiate the formation of rock glaciers (Humlum et al., 2007).

The active layer is the transition between the ground surface and a permafrost body and its chemical and physical properties can therefore influence interactions and exchange processes between these two entities. At the bottom of the active layer, the frost table prevents the infiltration of meltwater into the frozen parts of the subsurface. Although the frost table can be water-permeable as well (Ikeda et al., 2008), its shape can determine the direction and intensity of the flow of subsurface water (Hayashi et al., 2007). A particularly ice-rich frost table can prevent a thickening of the active layer, although exceptionally warm temperatures, like they were observed in the year 2003, can substantially damage this protective layer (Zenklusen Mutter and Phillips, 2012).

1.3 Overview of Periglacial Landforms

As the previous sections emphasized the importance of an enhanced understanding of the internal structure of periglacial landforms, the following section reviews the current state of research on the internal structure of selected types of periglacial landforms. This comprises rock glaciers, solifluction lobes, palsas/lithalsas and patterned ground, as investigations from these four landform types are presented in this thesis. Additionally, the methodological development of the applied geophysical methods is outlined.

Methodological Progress

Electrical Resistivity Imaging (ERI) and Ground-Penetrating Radar (GPR) are long-established geophysical methods for investigating the internal structure of landform units in periglacial environments (Doolittle et al., 1992; Farbrot et al., 2007; Isaksen et al., 2000; Kristensen et al., 2009; Ross et al., 2007). They are frequently used together, sometimes further supported by Seismic Refraction Tomography (SRT), to overcome the specific weaknesses of the single methods (Dusik et al., 2015; Kneisel et al., 2008; Otto and Sass, 2006).

One of the first applications of electrical resistivity measurements in permafrost-related research was presented by Østrem (1964), but the number of studies increased only from the 1980s onward. Basic 1-D approaches from that time enable a general detection of permafrost conditions, an assessment of the subsurface layering and estimations on the thickness of ground ice occurrences. Numerous publications show the worldwide application of 1-D resistivity measurements (Evin and Fabre, 1990; Fukuda and Sone, 1992; Haeberli, 1985; Ishikawa and Hirakawa, 2000; King and Seppälä, 1987; Kneisel, 1999; Kneisel et al., 2000; Osterkamp et al., 1980). Around the turn of the millennia, 1-D approaches were replaced by 2-D ERI. This two-dimensional approach permits a differentiation between areas with different subsurface conditions on a small distance and is suitable for more complex research issues, like detailed investigations of the internal structure of landforms (Hauck and Vonder

Mühll, 2003; Hauck et al., 2003; Kneisel, 2006). Repeatedly performed surveying in the same area permits the detection and investigation of processes on different timescales, like the percolation of meltwater or thawing of the active layer (Hauck, 2002; Hilbich et al., 2011; Hilbich et al., 2009; Kneisel et al., 2014b). This approach still represents the current state-of-the art.

Only in recent years, 3-D ERI was introduced into permafrost-related research (Kneisel et al., 2014a; Rödder and Kneisel, 2012b; Seppi et al., 2015). Through the ability to provide subsurface information on two horizontal dimensions, and in contrast to intersecting networks of two-dimensional models (Langston et al., 2011; Scapozza and Laigre, 2014), 3-D ERI allows to investigate spatial correlations between horizontally distributed surface and subsurface parameters. These spatial analyses can be used, e.g., to detect relationships between the occurrence of certain soil types and the horizontal permafrost distribution (Kneisel et al., 2015). Other studies use 3-D ERI for investigating the internal structure of solifluction lobes (Draebing, 2016) or use 3-D ERI models as a constraint for 3-D GPR (Merz et al., 2015). Investigations of the differences between 2-D ERI models and congruent slices of overlapping 3-D ERI models show that the magnitude of the modeled resistivity values and the extent of the detected structures can vary (Emmert and Kneisel, 2017).

In permafrost-related research, most of the resistivity data sets are processed with the widespread RES2DINV/RES3DINV software package (Geotomo Software Sdn Bhd). Despite the observation that results between different software packages can vary (Hellman et al., 2016), only a small number of studies use other software products (e.g., Emmert and Kneisel, 2015; Léger et al., 2017; Tran et al., 2018).

The application of GPR in permafrost-related research started from the 1970s onward (Annan and Davis, 1976). Similar 2-D GPR approaches as they were used back then are used still today. They are applied to investigate the layering of the subsurface, to assess frost table topography and to detect buried ice bodies (Brandt et al., 2007; Doolittle et al., 1992; Hinkel et al., 2001; Monnier et al., 2014; Schwamborn et al., 2008). Comparably to ERI, 3-D GPR approaches exist, permitting a three-dimensional assessment of the extent of subsurface features like ice-wedges or ice complex deposits (Munroe et al., 2007; Schennen et al., 2016). On the contrary, repeatedly performed monitoring approaches, e.g., to investigate processes in the active layer, are rare (Westermann et al., 2010).

Rock Glaciers

> *"Active rockglaciers [sic] are lobate or tongue-shaped bodies of perennially frozen unconsolidated material supersaturated with interstitial ice and ice lenses that move downslope or downvalley by creep as a consequence of the deformation of ice contained in them and which are, thus, features of cohesive flow"*
>
> — Barsch (1996).

The distribution of rock glaciers and their classification, especially in, but not restricted to remote areas, are in the focus of current research (e.g., Bolch and Gorbunov, 2014; Jones et al., 2018; Kenner and Magnusson, 2017; Knight et al., 2018; Schmid et al., 2015; Villarroel et al., 2018). In arid and semiarid areas, studies target the hydrologic significance of rock glaciers (Azócar and Brenning, 2010; Bodin et al., 2010). Estimations of the amount of water stored in rock glaciers, however, require additional information on the internal structure, in particular on ice content, which is commonly gained by geophysical methods (Bolch et al., 2018; Croce and Milana, 2002; Monnier and Kinnard, 2015).

The availability of long-term data from rock glaciers in Europe and North America enables investigations of the past and modeling of the future development of rock glaciers. Studies from the Swiss Alps show an initial acceleration of glacier movement at multiple rock glaciers around the 1990s (Roer et al., 2005) and a further acceleration after 1994 (Kääb et al., 2007). A strong increase in rock glacier movement is expected in the future, as models show that a temperature rise of 1 °C 10 a^{-1} can triple the acceleration (Müller et al., 2016). To affirm such projections, an enhanced understanding of subsurface processes, e.g., from modeling approaches (Pruessner et al., 2018; Scherler et al., 2014), and thus a detailed knowledge on the internal structure is crucial. However, it must be noted that results of previous studies indicate a predominant influence of external factors on rock glacier movement, compared to the influence of internal characteristics (Delaloye et al., 2008).

It has not yet been clarified whether the currently warming air temperatures already alter active layer thicknesses. At Murtèl rock glacier, borehole temperature records indicate an increase from 3.1 m to 3.5 m between 1987 and 2006 (Harris et al., 2009). However, data from another borehole on the same rock glacier between 2002 and 2010 does not affirm this assumption (Schneider et al., 2012). An increase is also not indicated by data from ten boreholes at different rock glaciers in the Swiss Alps between 1996 and 2009 (Zenklusen Mutter and Phillips, 2012). Intraannual active layer variations can be investigated by repeatedly performed 2-D ERI surveying: results from Murtèl rock glacier show a spatially variable thickening during the snowmelt period (Hilbich et al., 2009) and rapid and fundamental changes in the resistivity pattern at a smaller rock glacier in the same area (Kneisel et al., 2014b).

The occurrence of massive ice bodies in the root zones of rock glaciers and near the fronts has been detected and investigated by both GPR and ERI on many rock glaciers worldwide (Berthling et al., 2000; Degenhardt, 2009; Isaksen et al., 2000; Monnier et al., 2013; Ribolini et al., 2007; Ribolini et al., 2010). These ice patches are commonly assumed to represent (i) sedimentary ice/ice of sedimentary origin (i.e., remnant glacier ice or buried patches of snow or ice), or (ii) congelation ice (i.e., ice formed in the subsurface by freezing of liquid water) (Haeberli and Vonder Mühll, 1996). However, ground ice of both types can exist in close proximity (Kneisel and Kääb, 2007). Knowledge on the origin of ground ice allows to draw conclusions on the formation of rock glaciers, which is often assumed to be a sequentially process that is connected to glacier advances and retreats (Dusik et al., 2015; Monnier et al., 2008; Monnier et al., 2011; Monnier et al., 2013; Seppi et al., 2015).

The detection of ice content variations in the subsurface of rock glaciers provides information on hydrological conditions and flow paths (Hausmann et al., 2007; Hausmann et al., 2012; Leopold et al., 2008; Leopold et al., 2011). It further allows to draw conclusions on phenomena like flow-induced permafrost thickening and active layer thickening (Emmert and Kneisel, 2017; Kneisel, 2006). Differences in ice content between different rock glaciers are frequently attributed to differences in grain size, as fine-grained debris can hold only relatively low volumes of ice (Ikeda and Matsuoka, 2006; Kneisel, 2010; Schneider et al., 2013). To emphasize that a rock glacier is composed of materials with an extraordinary small grain-size, the term *pebbly rock glacier* is used (Ikeda and Matsuoka, 2006). This landform type is set in contrast to *bouldery rock glaciers* and describes landform units with a dominant clast size between 0.15 m and 0.2 m (Ikeda and Matsuoka, 2006).

Solifluction Lobes

> *"Isolated, tongue-shaped solifluction feature, up to 25 m wide and 150 m or more long, formed by more rapid solifluction on certain sections of a slope showing variations in gradient"*
>
> — Van Everdingen (2005).

This definition of solifluction lobes is based on the formative process of solifluction, which can include up to four components: (i) needle ice creep, (ii) frost creep, (iii) gelifluction and (iv) plug-like deformation (Ballantyne and Murton, 2017; Matsuoka, 2001). It further shows that solifluction is not bound to permafrost conditions or cold climates (French, 2018).

Current research focuses on the relationships between the different processes that are involved in the development of the landform units. This can be achieved by the measurement of surface and subsurface variables like e.g., temperature, displacement rate, soil moisture or pore water pressure (Harris et al., 2008a; Kinnard and Lewkowicz, 2005; Matsumoto and Ishikawa, 2002). Information on the type and speed of displacement is gained by photogrammetric approaches or laboratory experiments (Harris et al., 2003; Harris et al., 2008b; Matsuoka, 2014). Matsuoka (2001) presents schematic velocity profiles to connect vertical displacement gradients with the occurrence of certain subsurface processes. A morphometric classification system for a differentiation between solifluction lobes and similarly shaped small pebbly rock glaciers is presented by Matsuoka et al. (2005), who use the ratios between width, length and height to distinguish between the different landform types.

To understand the distribution of solifluction lobes on alpine talus slopes, detailed information on the internal structure of the hosting slopes is crucial. Affected talus slopes are known to show isolated aggregations of subsurface ice as well as interbedded ice-rich layers (Lambiel and Pieracci, 2008; Scapozza et al., 2011). Values of the surface and subsurface temperature show that ground thermal regimes at talus slopes with solifluction lobes are strongly influenced by spatial variations of snow cover (Luetschg et al., 2004). The aforementioned studies show a zonation of the permafrost probability: The probability for the occurrence of permafrost is higher in the lower parts of talus slopes, while it is lower in the upper parts. It is assumed that this pattern is caused by air circulations (Delaloye and Lambiel, 2005) or by avalanche snow deposits (Kenner et al., 2017).

The relationship between vegetation and solifluction processes is investigated by Eichel et al. (2016). Eichel et al. (2017) highlight the role of the pioneer plant *Dryas octopetala* as an engineering species on decelerating the displacement rates of solifluction lobes. Further environmental parameters that influence solifluction are snow cover (Hugenholtz and Lewkowicz, 2002; Ridefelt and Boelhouwers, 2006) and slope angle (Benedict, 1970a). Results of 2-D ERI on a solifluction slope in northern Sweden show that the formation of solifluction lobes can be associated with recent permafrost conditions in the subsurface, but that this is not a prerequisite (Kneisel, 2006; Kneisel, 2010). At lobes on a solifluction slope in Iceland, results of the same approach by Kneisel et al. (2007) show a shallow active layer and that the lobes are connected to recent permafrost conditions. The application of 3-D ERI on a solifluction lobe in the Swiss Alps permitted the delimitation of the three-dimensional geometry of a lobe body by Draebing and Eichel (2017).

Palsas/Lithalsas

Until today, there is no commonly accepted definition of the term *palsa* (Pissart, 2002; Seppälä, 1972; Seppälä, 2011; Washburn, 1979) but the occurrences of segregation ice and peat are considered to be the main characteristics of a frost mount to be classified as a palsa (Van Everdingen, 2005). As similar frost mounds also exist in settings without larger amounts of peat, Pissart (2002) suggests to use the term *lithalsa* to distinguish between the two landform types. However, this differentiation is not yet prevalent, and therefore landform units with and without peat are regarded as palsas in this thesis.

The cyclic development of palsas and the formation and degradation of palsa landscapes are main objectives of current research. Investigations comprise the analysis of vegetation patterns and soil layering, as well as dating approaches based on plant macrofossils or tephra layers (Hirakawa, 1986; Kuhry, 2008; Matthews et al., 1997; Oksanen, 2006; Zuidhoff and Kolstrup, 2005). Since the complex energy fluxes between the surface and the subsurface are not fully understood yet (Sjöberg et al., 2016), the interest in the response of palsas to environmental changes is particularly high. Studies from several regions document a decline in palsa landscapes: This decline was around 50 % in a study region in Sweden between 1960 and 1997 (Zuidhoff and Kolstrup, 2000) and ranged between 33 % and 71 % in different study regions in Norway between the 1950s and today (Borge et al., 2017). In subarctic palsa mires in Sweden, the increase in active layer thickness between 1978 and 2006 was between $0.7\,\mathrm{cm\,a^{-1}}$ and $1.3\,\mathrm{cm\,a^{-1}}$ at different palsas (Åkerman and Johansson, 2008). An increasing active layer thickness is also reported from Orravatnsrústir palsa site in central Iceland (Saemundsson et al., 2012), which is one of the study sites of this thesis (see sect. 2.2).

In contrast to direct observations of the internal structure (Allard and Rousseau, 1999; Iwahana et al., 2012), geophysical approaches are non-destructive. Main targets for GPR surveying are assessments of the shape of the frozen core and the thickness of the active layer (Doolittle et al., 1992; Kohout et al., 2014). The application of 2-D ERI has also proven its suitability for investigations of the internal structure of palsas, in particular for assessing ice content variations and for estimating the depth of the frozen layer (Dobiński, 2010; Fortier et al., 2008; Kneisel, 2010; Kneisel et al., 2007; Lewkowicz et al., 2011; Wolfe et al., 2014). For a detailed understanding, a joint application of both methods was performed by Sjöberg et al. (2015). Findings of the aforementioned case studies agree on the occurrence of small-scale areas of frozen and unfrozen conditions and show that the thickness of the active layer is spatially variable.

Patterned Ground

> *"The term patterned ground is used to describe terrain that exhibits regular or irregular surface patterning, most commonly in the form of circles, polygons, irregular networks or stripes"*
>
> — Ballantyne and Murton (2017).

This definition includes several subtypes, such as sorted or non-sorted forms, and various formative processes, like differential frost heave or frost cracking. It further includes frost-independent processes like mass displacements or desiccation (Washburn, 1979). Nevertheless, intense seasonal frost is included in other definitions (e.g., Washburn, 1956) and most patterned ground phenomena are attributed to repeated freeze-thaw cycles (Ballantyne and Murton, 2017).

Current research, focusing on patterned ground include (i) the relationships between vegetation and morphology (Walker et al., 2004), (ii) formative processes (Matsuoka et al., 2003; Peterson and Krantz, 2008), and (iii) interactions between the occurrence of patterned ground and other variables, like e.g., soil texture (Feuillet et al., 2012; Luoto and Hjort, 2006; Watanabe et al., 2017). An inconsistent development of patterned ground occurrences was observed in the Canadian Arctic between 1972 and 2004 (Steedman et al., 2017). This observation indicates that interactions between patterned ground and warming temperatures are complex and that a strong influence of subsurface properties must be considered. Despite the low number of case studies in which geophysical methods are used for investigating the internal structure of patterned ground, they are assumed to be valuable tools, especially for investigating formative processes within the active layer (Kasprzak, 2015; Kneisel, 2006).

As the definition of patterned ground includes a variety of surface phenomena, case studies show varying subsurface conditions between different types of patterned ground. Results of a joint application of 2-D ERI, 2-D GPR, Electromagnetic techniques and LiDAR in the Canadian Arctic show different distribution patterns of the addressed geophysical variables between different types of polygons (Hubbard et al., 2013). For patterned ground in Svalbard, results of 2-D ERI show different resistivity distribution patterns between a site with ice-wedge polygons and a site with surface cracks on mudboils (Watanabe et al., 2012). Both of the aforementioned studies attribute the observed differences between the different types of patterned ground to differences in surface parameters like soil moisture, soil temperature or dominant soil texture.

At a study site with sorted stone circles in Svalbard, results of 2-D ERI show an undulating frost table topography and indicate intense cryoturbation processes. This points towards a close connection between frost table and surface topography (Kasprzak, 2015). In contrast to the undulating frost table topography in the subsurface of the sorted stone circles of the aforementioned study, a rather planar frost table topography was detected in the subsurface of non-sorted polygons in Svalbard and in the subsurface of frost-crack polygons and sorted polygons in Iceland (HPA study site of this thesis) (Kasprzak, 2015; Kneisel, 2010; Kneisel et al., 2007; Watanabe et al., 2012).

1.4 Objectives

After the importance of an enhanced knowledge on the internal structure of periglacial landforms was emphasized in the previous sections, this section focuses on the objectives of this thesis, stated as research questions. They are divided into three groups: Questions of group 1 address the spatial domain, i.e., the composition of the landform units. This is a basic requirement for the following questions of group 2. Questions of this group focus on the process-domain and thus add a temporal perspective. Targeted changes represent either long-term adaptions to changing environmental conditions or responses to short-term or seasonal variations. Questions of group 3 focus on the suitability of the applied methods for answering the questions of the first two groups and address the performance of the 3-D ERI approach.

Group 1: Composition of the Investigated Landform Units

- Are the investigated study sites/landform units currently affected by permafrost conditions?
- To what extent do spatial variations of the internal structure (e.g., active layer thickness, frost table topography, ice content) occur?
- What is the origin of the detected ground ice?
- Which prominent resistivity structures or GPR reflection patterns appear in the results of different landform units or landform types?
- Does the spatial distribution of ground ice correspond to the spatial distribution of surface parameters, such as surface topography or vegetation?

Group 2: Permafrost and Landform Development

- How does the internal structure interact with geomorphologic processes?
- Which processes determined landform development in past phases of colder climate?
- How will warming temperatures alter the permafrost distribution and affect the landform development?

Group 3: Methodological Approach

- Are the applied methods suitable to answer the research questions?
- Which advantages provides the relatively innovative and until today sparely used 3-D ERI approach over the established 2-D ERI approaches? Which limitations exist and how can the 3-D ERI approach be further developed?
- Which differences occur between resistivity models generated with different inversion software?

Chapter 2
Study Sites

This thesis comprises findings from six study sites (NAR, UER, LTF, FUR, OVR and HPA), located in high-alpine (fig. 2.1a, b) or subarctic areas (fig. 2.1c), to reflect a wide range of permafrost environments. The sites were chosen with respect to logistical aspects and accessibility, but the availability of previous studies was also considered, as this permits an integration of the results into a wider context. Each study site comprises one or more landform units of different or the same landform types. This chapter presents an overview of the main topographic, geological and climatological characteristics of the study sites. While site-specific details are noted in individual sections (2.1–2.6), general characteristics are presented at the beginning of this chapter.

High-Alpine Sites

Strong relief intensity characterizes high-alpine permafrost environments. Elevations of the investigated landform units are between 2460 m a.s.l. and 2840 m a.s.l. and all landform units are located in north-exposed positions. Long-term average temperature and precipitation records (fig. 2.1d, e) indicate a *DfC* climate at the high-alpine sites, following the Köppen-Geiger classification system (Peel et al., 2007). The study sites NAR, UER and LTF are in the Swiss canton of Grisons and are part of, or adjacent to the Engadin valley, a 130 km long valley drained by the river Inn. Data from the weather station in Samedan (fig. 2.1d) shows the dry climatic characteristics of an inner-alpine valley and thereby contrasts the rather moist conditions that are observed near study site FUR in central Switzerland.

The MAAT in Samedan between 1961 and 1990 was 1.3 °C (MeteoSchweiz, 2018). When a value of 0.56 °C 100 m^{-1} is assumed for the non-adiabatic temperature gradient (Gensler, 1978), the calculated position of the −2 °C isotherm is around 2300 m a.s.l. Thus, all investigated landform units of NAR, UER and LTF are located in the part of the periglacial zone where frost-action conditions dominate (French, 2018). Due to the rather dry climatic conditions, only relatively small areas of the Engadin are glaciated today. The higher precipitation values that are recorded near FUR study site (fig. 2.1e), explain the more extensive glaciation in the surrounding area and the high rate of avalanche events that is documented in newspaper articles (e.g., Felder, 2018).

At the Alpine Research Station Furka (ALPFOR) (2440 m a.s.l., horizontal distance to the investigated area is around 800 m), temperature records show an MAAT of 0.13 °C between 2013 and 2015 and a mean annual precipitation of 1181 mm (Hiltbrunner, 2017, unpublished data). This aligns with records from the weather station in Gütsch ob Andermatt (fig. 2.1e). Assuming a value of 0.56 °C 100 m^{-1} for the non-adiabatic temperature gradient (Gensler, 1978), and using the long-term MAAT between 1981 and 2010, the 2 °C isotherm is located at an elevation of 2712 m a.s.l. This elevation is higher than the elevation of the investigated

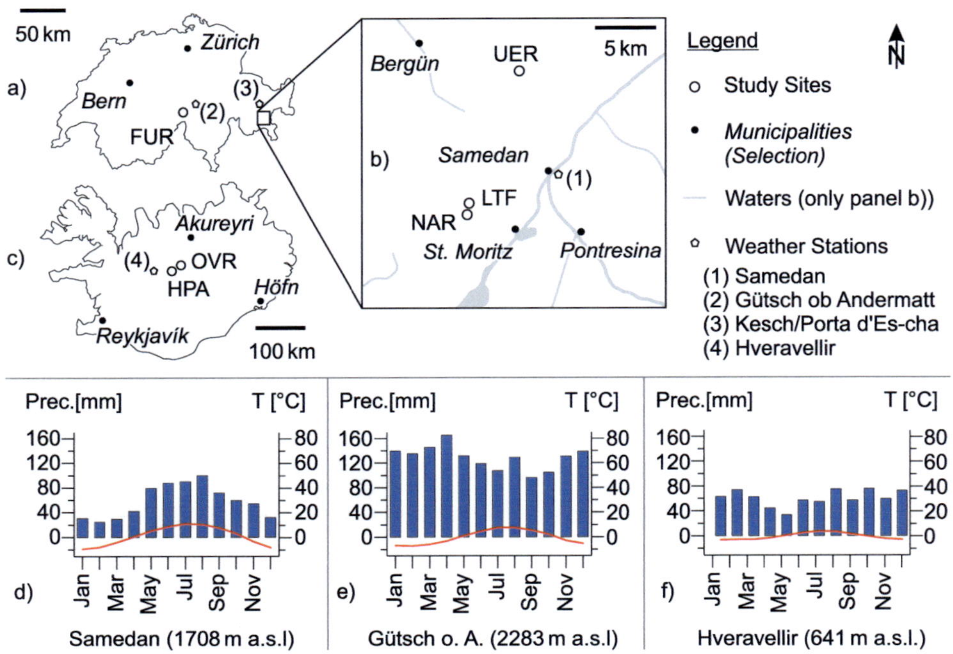

Figure 2.1: Study Sites. a) High-alpine Sites, b) Engadin Sites, c) Subarctic Sites, d–f) Climate Diagrams. Data from MeteoSchweiz (2018) and Icelandic Met Office (2018), respectively.

landform units in this area, which are thus, in contrast to the landform units of the other high-alpine study sites, not necessarily dominated by frost-action conditions (French, 2018). From a tectonic view, NAR, UER and LTF are located in an area that belongs to the unit of the Lower Austroalpine Nappes. These nappes represent the former Adriatic continental margin and include a crystalline Pre-Triassic basement below Mesozoic sedimentary rocks (Pfiffner, 2015). Study site FUR is located in the Urseren Zone, where a small band of Mesozoic rocks is exposed between the crystalline Aar and Gotthard massifs (Pfiffner, 2015).

Subarctic Sites

The two subarctic study sites OVR and HPA are located in the uninhabited and mostly barren desert of central Iceland, from which detailed climatic data is not available. The closest weather station at Hveravellir is located around 55 km from OVR and around 40 km from HPA. Temperature and precipitation records from this weather station (fig. 2.1f) show a cold tundra climate, specified as ET in the Köppen-Geiger classification system (Peel et al., 2007). Geological maps show that OVR is in a zone of basic and intermediate interglacial and supraglacial lavas with intercalated sediments that are younger than 0.8 Ma (Jóhannesson, 2014). Study site HPA is located in a zone of basic and intermediate hyaloclastites, lava and associated sediments of a similar age, but geological maps show the additional occurrence of prehistoric lava flows that are older than AD 871 (Jóhannesson, 2014).

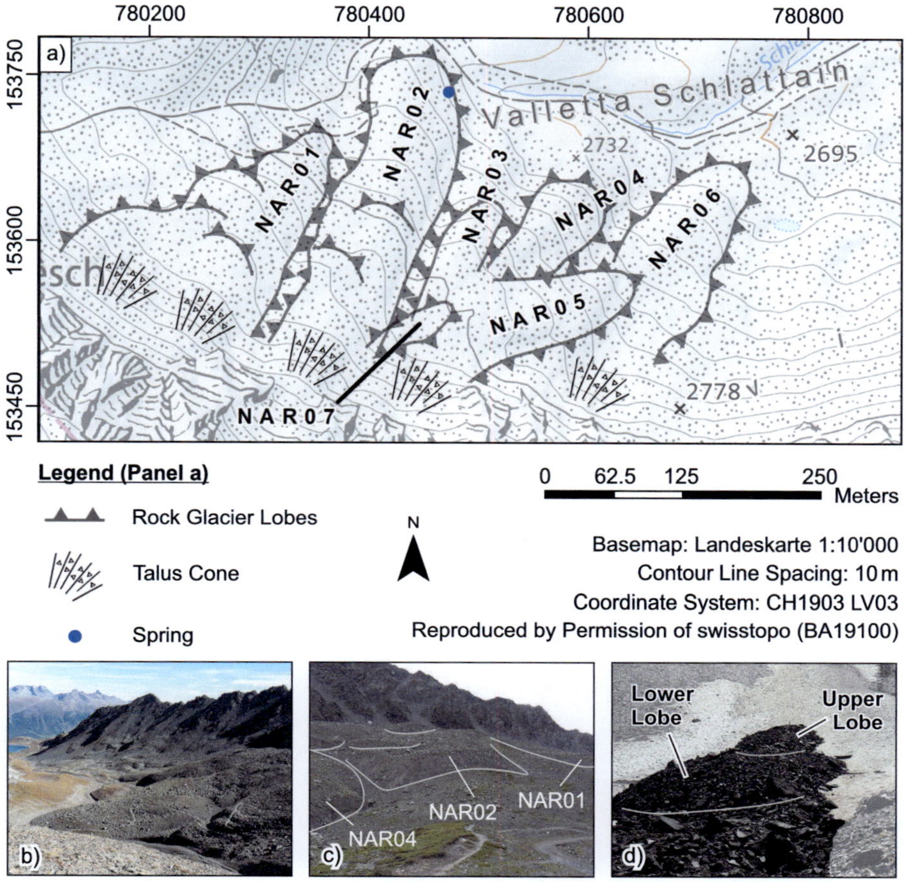

Figure 2.2: NAR Study Site. a) Overview Map, b) Overview Photo, c) Photo NAR04, NAR02 and NAR01, d) Photo NAR07. Own Pictures.

2.1 Piz Nair Rock Glacier Assembly (NAR)

Study site NAR (46.52° N, 9.78° E), is administratively assigned to the municipality of Celerina and located in the 2 km long, east-striking valley *Valletta Schlattain*, which is drained by the headwaters of a small creek. Main target of investigations at NAR is an assembly of north- or northeast-striking rock glaciers below the summit of Piz Nair and the adjacent mountain ridge (fig. 2.2). The rock glaciers vary in length and width and partially have a multi-lobe appearance (see tab. 2.1). The composition of the talus materials at the surface of the rock glaciers reflects the lithology of the bedrock exposed in the summit area of Piz Nair and includes mainly shales and breccias of Mesozoic age from the Err-Nappe (Peters et al., 2005). The clast size of all rock glacier lobes is small, and the rock glaciers are hence classified as pebbly rock glaciers.

Table 2.1: Overview of Investigated Rock Glaciers NAR.

Label	Length [m]	Width [m]	Comment
NAR01	190	120	Two Lobes
NAR02	270	80	Three Lobes
NAR03	130	40	Joint Root Zone with NAR05
NAR04	100	90	Surface Cracks
NAR05	150	60	Joint Root Zone with NAR03
NAR06	120	100	Surface Cracks
NAR07	60	20	Two Lobes, In Root Zone of NAR03/NAR05

Ancient topographical maps indicate the existence of surface ice between 1917 and 1944 (Coaz et al., 1925; Coaz et al., 1946), but today the valley is completely ice-free at the surface. The occurrence of ground ice at the summit of Piz Nair is known at least since the construction of a cable car station in the 1950s (Haeberli, 1992). Today, the Alpine Permafrost Index Map (APIM) (Boeckli et al., 2012) describes the spatial distribution of permafrost in most parts of the investigated area with colors that indicate "permafrost in nearly all conditions" and only small patches at the valley floor show colors that represent a medium or low potential for the occurrence of permafrost. Similar conditions are indicated by the Swiss Map of Potential Permafrost Distribution (BAFU, 2005), which shows colors that represent the attribute "Extensive permafrost likely" throughout the complete investigated area. Around the positions of the talus cones, it further states an "increasing thickness" of permafrost.

A morphological description of the rock glacier assembly and results of geophysical surveying on multiple rock glaciers with different methods is presented by Ikeda and Matsuoka (2006). Their results indicate permafrost conditions at the rock glaciers NAR01 and NAR06. At NAR01, subsurface temperature records show that the upper boundary of a permafrost layer is located between 3 m and 5 m depth (Emmert and Kneisel, 2017).

2.2 Rock Glaciers Piz Üertsch and Piz Blaisun (UER)

Study site UER (46.60° N, 9.84° E) is located in the 5 km long, north-striking valley *Val Plazbi*, which administratively belongs to the municipality of Bergün. At the head of the valley, an extensive glaciation during the Little Ice Age (LIA), as depicted on ancient topographical maps (Coaz and Leuzinger, 1878), is outlined by lateral moraines. However, the recent extent of surface ice is restricted to a small ice patch in a cirque (fig. 2.3a). The valley further comprises multiple rock glaciers, of which two are targeted by investigations presented in this thesis: UER01 and UER02. Both rock glaciers consist of fine-grained debris of shale and marlstone from the Ela-Nappe, which is exposed at the mountain ridge between Piz Üertsch and Piz Blaisun. Due to the small grain size of the clasts, both rock glaciers are classified as pebbly rock glaciers. Only at the summit area of Piz Üertsch, a shallow layer of carbonate rocks exists (Bearth et al., 1987).

According to the APIM (Boeckli et al., 2012), the occurrence of permafrost is highly probable only for the uppermost part of UER01, while UER02 is completely within the zone of maximum permafrost probability. This disagrees with the depiction on the Swiss Map of

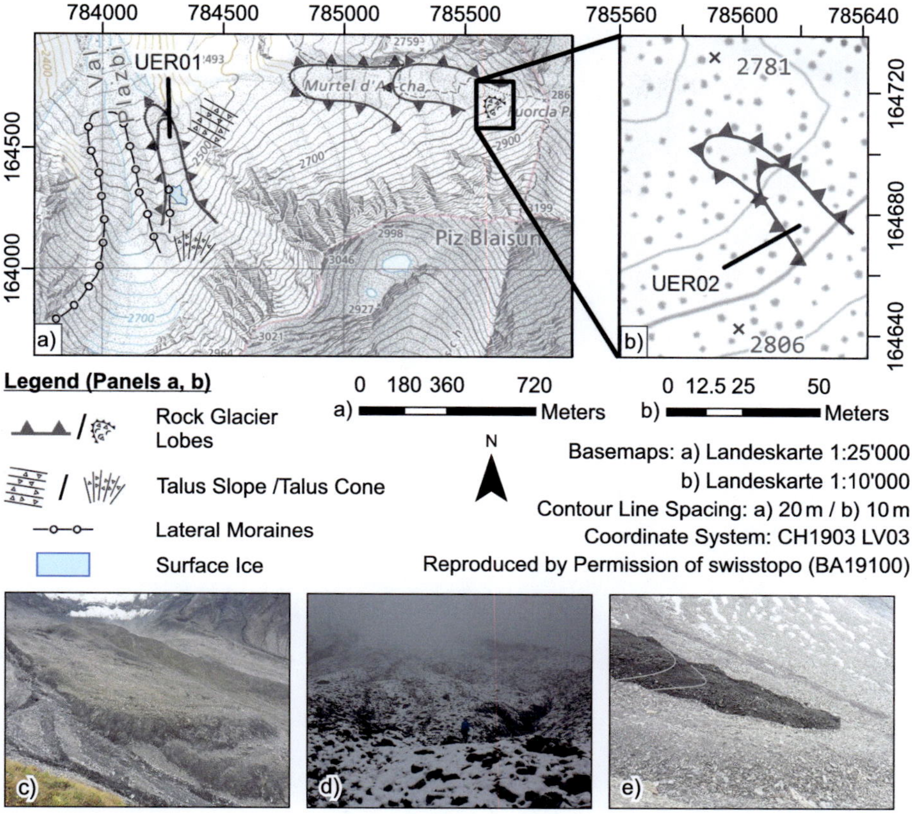

Figure 2.3: UER Study Site. a), b) Overview Maps, c) UER01 and Adjacent Lateral Moraines, d) Furrow-and-Ridge Topography of UER01 Rock Glacier Snout, e) UER02. Own Pictures.

Potential Permafrost Distribution (BAFU, 2005), according to which extensive permafrost also affects the rock glacier snout of UER01.

Rock glacier UER01 (fig. 2.3a, c–d) is around 500 m long and 120 m wide. It parallels the LIA moraines by around 100 m. In front of the rock glacier, sunken and hence apparently relict lobes indicate a successive rock glacier formation (fig. 2.3c). The elevation of UER01 ranges from 2394 m a.s.l. at the rock glacier front to 2554 m a.s.l. at the root zone. It is divided into three zones: The root zone (i) comprises a perennial surface ice patch of around 2400 m^2. A small lateral moraine on the western edge indicates that the glaciation on the rock glacier surface was more extensive in the past, as depicted on ancient topographical maps (Coaz and Leuzinger, 1878). The central zone of the rock glacier (ii) comprises longitudinal surface ridges of around 0.5 m to 1 m height in the eastern and central part, as well as a prominent surface depression in the western part. The rock glacier snout (iii) comprises arcuate surface structures on the surface, contoured by deep furrows (fig. 2.3d). This marked furrow-and-ridge topography, which reaches a height difference of up to 4 m, indicates that strong flow processes were active in the past (Frehner et al., 2015; Kääb and Weber, 2004).

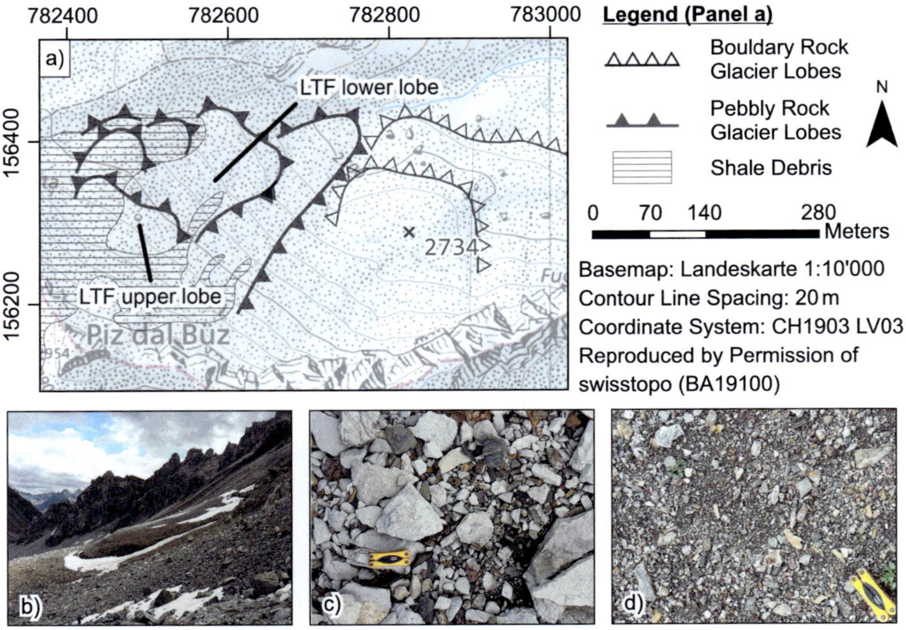

Figure 2.4: LTF Study Site. a) Overview Map, b) View on Upper Lobe, c) Coarse-grained Section, d) Fine-grained Section. Own Pictures.

The second target of investigations, UER02, is a small, around 50 m long northwest-striking rock glacier that consist of multiple lobes (fig. 2.3b, e). It is located 1.3 km east of UER01 at an elevation of around 2790 m a.s.l. on the western slope of Piz Blaisun. Morphometrical analyses that include assessments of length, slope, aspect and lithology of UER01 rock glacier are presented in a study by Ikeda and Matsuoka (2006), but no results of geophysical surveying are published therein. Emmert and Kneisel (2017) validate the occurrence of permafrost by subsurface temperature records and present results of 3-D ERI and 2-D SRT as well as a geomorphological map of the rock glacier snout and in the central part of UER01.

2.3 Las Trais Fluors Rock Glacier (LTF)

Only 3.5 km northeast of NAR, LTF study site (46.53° N, 9.82° E) is located within the boundaries of the municipality of Samedan. The investigated area includes a section of a rock glacier assembly on the northern slope of the *Las Trais Fluors* mountain ridge that includes both pebbly and bouldery rock glaciers (fig. 2.4a). Like at NAR, ancient topographical maps show that a small patch of surface ice existed between 1917 and 1944 (Coaz et al., 1925; Coaz et al., 1946), but today the surface of the area is ice-free.

Target of investigations at LTF is a multi-lobe rock glacier at the northern slope of the mountain Piz dal Büz, named LTF rock glacier in this thesis. The upper lobe is around 70 m long and 110 m wide (fig. 2.4b) and the lower lobe is around 120 m long and 90 m wide. In contrast to the upper lobe, which has a distinct outline, the outline of the lower lobe is more complex and subdivided at its northwestern margin. The elevation of the rock glacier is between 2835 m a.s.l. at the root zone of the upper lobe and 2744 m a.s.l. at the front of the lower lobe. Different rock types of the Err-Nappe are exposed in the summit area of Piz dal Büz, and hence the rock glacier lobes below are longitudinally divided into two segments that comprise different debris materials: relatively coarse-grained dolomite debris of white color (fig. 2.4c) covers the eastern part of the lobes, while relatively fine-grained brown shale debris covers the western part of the lobes (fig. 2.4d).

Both investigated rock glacier lobes lack a distinct furrow-and-ridge surface topography and are classified as pebbly rock glaciers, despite the relatively coarse clast size of the dolomitic debris (Ikeda and Matsuoka, 2006). Maps of the potential permafrost distribution like APIM (Boeckli et al., 2012) or the Swiss Map of Potential Permafrost Distribution (BAFU, 2005) agree on a maximum level of permafrost probability throughout the complete investigated area. Study site LTF is well-known for investigations of frost weathering and rockwall erosion (Matsuoka, 2008), but also investigations of the morphometry of the rock glacier lobes and their velocity were performed. Permafrost conditions below a depth of 3 m were confirmed at the upper lobe by borehole temperature measurements between the years 2000 and 2005 (Ikeda et al., 2008). Results of previous studies show that the upper and the lower lobe of LTF rock glacier are in an active state (Ikeda and Matsuoka, 2002). Very fast deformation rates with averages of up to 1 $m\,a^{-1}$ and resistivity values of around 1 kΩm for frozen materials indicate the existence of a water-permeable permafrost table (Ikeda, 2006; Ikeda and Matsuoka, 2006; Ikeda et al., 2008). According to direct observations, the shallow subsurface of the upper lobe is ice-cemented and borehole temperature records show that subzero temperatures are present below a depth of 3 m (Ikeda and Matsuoka, 2006).

2.4 Blauberg/Furka Pass (FUR)

Study site FUR (46.57° N, 8.42° E) is located on the border between the municipalities of Realp and Obergoms and comprises the northern slope of the mountain Blauberg, close to Furka Pass (fig. 2.5a, c). Numerous bound and unbound lobe-shaped structures occur on the surface of the entire slope between 2745 m a.s.l and 2430 m a.s.l. The lobes are built up of metamorphic debris from the summit area, which comprises mostly gneiss, schist and phyllites of Paleozoic age (Flück et al., 1975; Labhart et al., 2012). Despite the partially extensive glaciations in the surrounding area, the investigated area is free of surface ice today. According to maps of the potential permafrost distribution, the occurrence of extensive permafrost is highly probable on the entire slope, with increasing thickness in an upslope direction (BAFU, 2005; Boeckli et al., 2012).

Bound lobes concentrate at elevations of below 2500 m a.s.l. and are textbook examples for solifluction lobes (Stahr and Langenscheidt, 2015). At elevations above 2500 m a.s.l., the lobes are predominantly unbound. In addition to the hosting slope, two lobe-shaped structures of the upper section of the slope are the further targets of investigation at FUR (fig. 2.5b): the upper lobe FUR01 is around 1.5 m high and a rather isolated landform unit (fig. 2.5d), while the lower lobe FUR02 (fig. 2.5e) is part of a compound structure of several lobes and around 1.8 m high. To the knowledge of the author, no geomorphological or other permafrost-related studies from FUR are published.

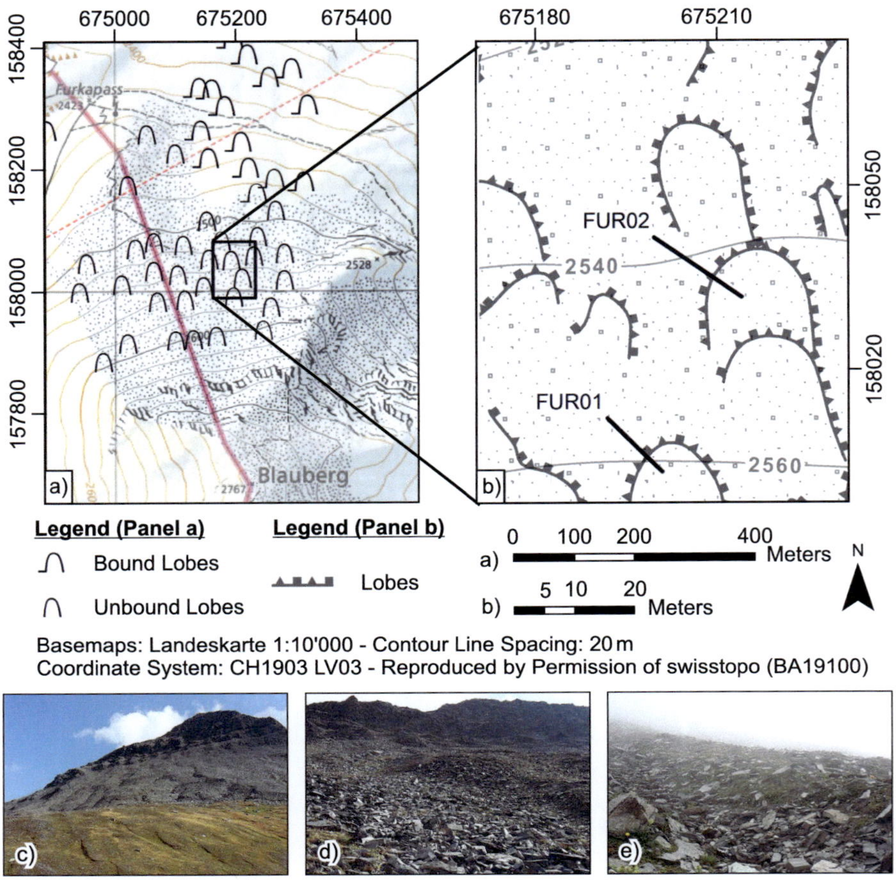

Figure 2.5: FUR Study Site. a), b) Overview Map, c) Overview Picture, d) Side View on FUR01, e) Side View on FUR02. Own Photos. Bold Red Line in Panel a) Indicates Cantonal Boundary.

2.5 Orravatnsrústir Palsa Site (OVR)

Study site OVR (65.08° N, 18.53° W), located 14 km north of Hofsjökull glacier, is part of the municipality Sveitarfélagid Skagafjördur. The study site stretches from a large wetland area at the south of lake Orravatn to the surrounding desert (fig. 2.6). While palsas within the wetland area are relatively small, palsas surrounded by water are markedly higher (see tab. 2.2). In contrast to the barren desert, the wetland area is densely vegetated with mosses and dwarf trees on the palsas while sedges and grasses dominate the main part of the wetland area. The formation of ground ice at OVR started not until after 4500 a before today (Hirakawa, 1986).

Targets of investigations at OVR study site are five distinct palsas of different height, size and shape (fig. 2.6b–e, tab. 2.2) and parts of their surrounding areas. As the palsas of OVR lack a massive peat layer, likely because the organic content is lowered by a steady aeolian deposition (Saemundsson et al., 2012), they are classified as lithalsas or lithalsa plateaus, fol-

Table 2.2: Overview of Investigated Palsas OVR.

Label	Palsa Height [m]	Extent of Uplifted Area [m^2]	Description
OVR01	0.5	30	Water Logs
OVR02	1	240	Uplifted Area
OVR03	2.4	1090	Part of Former Plateau, Surface Cracks
OVR04	2.4	1400	Part of Former Plateau, Surface Cracks
OVR05	1	40	Surface Crack

lowing the definition of Pissart (2002). An aerial picture from 1998, which is reprinted by Saemundsson et al. (2012), indicates that palsas OVR03 and OVR04 are remnants of one large palsa or palsa plateau. Ongoing lateral degradation was visually observed in the field by an enlargement of the gap between OVR03 and OVR04 between 2015 and 2017.

Results from previous studies at OVR show increasing thicknesses of the active layer. At OVR03, active layer thickness increased from 0.69 m in 2004 to 0.81 m in 2010 (Saemundsson et al., 2012). Results from Kneisel et al. (2007) and Kneisel (2010) show that the average active layer thickness increased from between 0.45 m and 0.65 m in 2001 to between 0.77 m and 0.81 m in 2006 and that the depth of the permafrost layer is between 5 m and 7.5 m. The MAGST between July 2006 and July 2007 was 1 °C, and the curve of daily mean temperatures indicates absence of a thick snow cover on the palsas (Kneisel, 2010). Despite the visually observed signs of degradation, Kneisel (2010) states that small palsas at OVR presumably represent an early development stage and that palsa formation is hence still possible under the recent environmental conditions.

2.6 Hofsjökull Proglacial Area (HPA)

Like OVR, HPA (64.98° N, 18.84° W) is part of the municipality of Sveitarfélagid Skagafjördur. The investigated area is at a distance of around 500 m from the northern margin of Hofsjökull glacier, which retreated by around this distance in historic times (Oddur et al., 2013). The investigated area comprises an extensive plain surface that is intersected by isolated exposures of lava (fig. 2.7a). The material composition of the plain surface consists of a mixture of coarse clast and fine-grained materials. At positions next to the lava exposures, larger clasts are absent and fine-grained materials accumulate.

Targets of investigations are two areas of patterned ground between the mountains Krókafell and Tvifell (HPA01, HPA02), which are separated by around 100 m (fig. 2.7a, d). At HPA01, sorted polygons with diameters between 1 m and 2 m occur on the plain surface, which comprises material of different grain size. (fig. 2.7b). At HPA02, smaller surface cracks with maximum diameters around 1 m occur at a spot where fine-grained materials accumulate (fig. 2.7c). Results of previous 2-D ERI surveys indicate permafrost conditions at HPA01 and that the subsurface consist of horizontally layered unconsolidated sediments (Kneisel, 2010). The occurrence of massive ground ice was detected in the subsurface of a nearby moraine structure, close to the mountain Krókafell (Kneisel et al., 2007). Temperature records at HPA01 show a MAGST of around 1 °C and indicate the formation of a relatively thick winterly snow cover, compared to OVR (Kneisel, 2010).

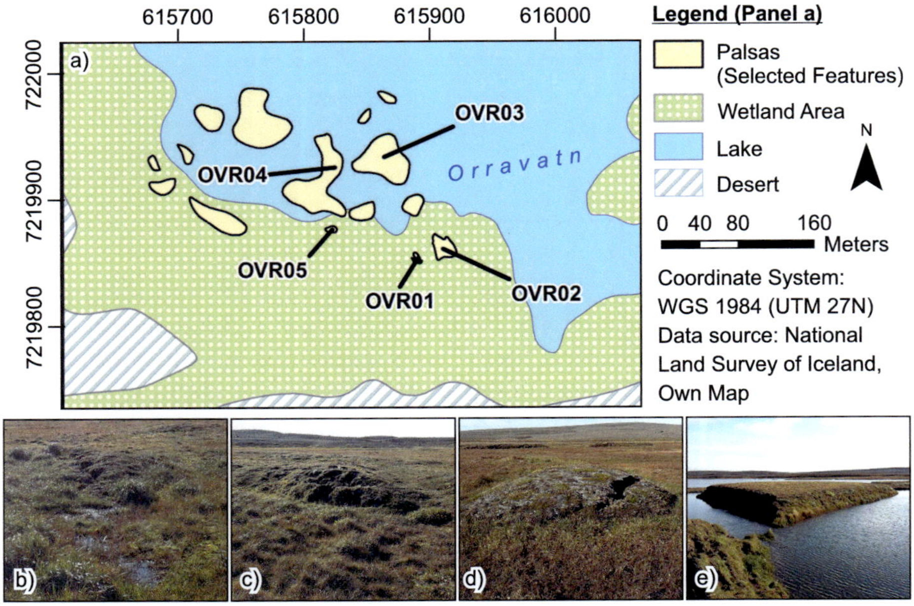

Figure 2.6: OVR Study Site. a) Overview Map, b) OVR01, c) OVR02, d) OVR05, e) OVR04 (front) and OVR03 (back). Own Photos.

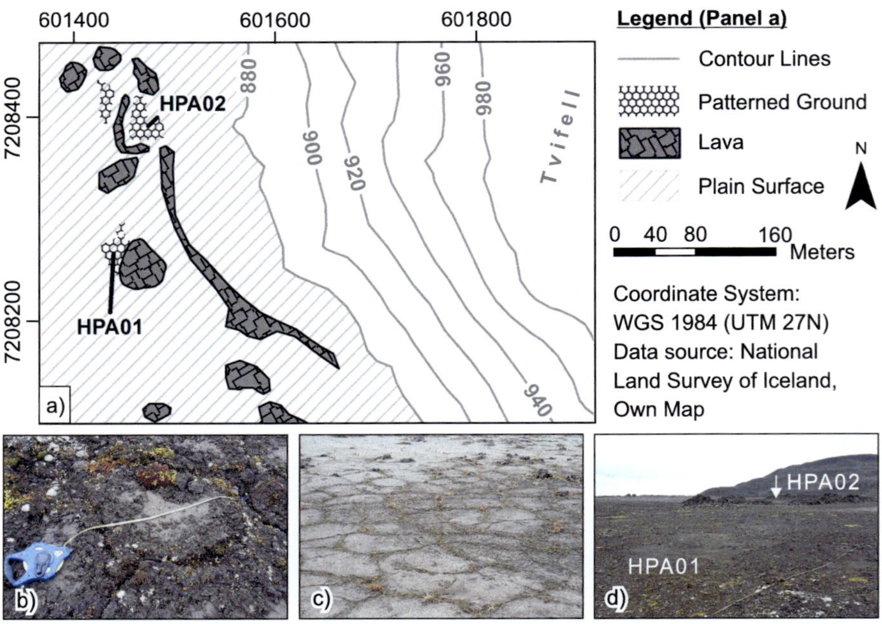

Figure 2.7: HPA Study Site. a) Overview Map, b) Sorted Polygons on Plain Surface, c) Surface Cracks in Fine-grained Materials, d) Overview Picture. Own Photos.

Chapter 3
Methods

This chapter describes the different methods that were used for the presented investigations. Beside the two geophysical methods ERI (3.1) and GPR (3.2), complementary measurements include the continuous recording of (i) surface and subsurface temperature values and the recording of (ii) soil matric potential values, used for assessments of the content of liquid water in the subsurface (3.3). At the end of the chapter, an overview of the performed surveys and their locations is attached (3.4).

Geophysical techniques provide information on the distribution of physical properties in the subsurface. This allows to conclude on the material composition and its characteristics. Together with the non-invasive way of data acquisition, this makes them excellent tools for various issues of engineering, exploration and environmental sciences. However, as most geophysical methods provide indirect information only, they do not provide one unique solution, but models that approximate the real conditions in the subsurface (Reynolds, 2011).

To improve the validity of geophysical models, joint applications of two or more geophysical methods are frequently applied, especially in situation where ground truth information is absent (see e.g., Briggs et al., 2017; Dusik et al., 2015; Sjöberg et al., 2015). These joint applications can compensate the weaknesses of a single method, as different geophysical methods are based on different physical parameters. Therefore, geophysical models are not only an additional source of information to direct observations, but can be seen as a "primary source of evidence" for different research issues (Van Dam, 2012).

The basic theory and data processing procedures of different geophysical methods are described in various textbooks (e.g., Lowrie, 2011; Reynolds, 2011; Telford et al., 1990). Their application in a geomorphologic context is outlined e.g., by Schrott and Sass (2008) or Van Dam (2012), and applications in permafrost-related research are described e.g., by Hauck (2013), Kneisel et al. (2008) or Scott et al. (1990).

3.1 Electrical Resistivity Imaging

The application of electrical resistivity methods is based on the varying electrical properties of different earth materials. Through its representation as resistivity models, which consist of discrete model cells, the subsurface resistivity distribution provides information on the structure and the composition of the subsurface. The spatial resolution of these resistivity models is based on the extent of the two- or three-dimensional model cells and can be adjusted to variations in the density of data points. Today, multiple approaches of data acquisition and processing techniques exist, with varying nomenclatures between different authors. Hence, it is necessary to state the nomenclature that is used in this thesis. This concerns the following notations:

- ERI/ERT: In this thesis, the abbreviation ERI is used for all types of resistivity surveys, regardless of dimensions. The term is preferred over ERT, as ERT implies that tomograms are derived as results, which is not the case for three-dimensional surveys. The abbreviation ERT is only used as a part of the abbreviation ERTM (Electrical Resistivity Tomography Monitoring) for repeatedly performed monitoring surveys, as this approach was solely performed two-dimensional in the presented investigations.

- 2-D/3-D: This prefix, used for resistivity surveys or models, refers to the parametrization of the inversion scheme only, regardless of the type of data acquisition. A two-dimensional inversion scheme permits the modeled resistivity distribution to vary in two dimensions, while a three-dimensional inversion scheme permits the modeled resistivity distribution to vary in three dimensions.

- r-3-D/q-3-D: To distinguish between the two main types of data acquisition for 3-D models, the prefixes r-3-D and q-3-D are used as abbreviations for *real-3-D* and *quasi-3-D*. While data acquisition by the r-3-D approach uses a rectangular grid of electrodes, data acquisition by the q-3-D approach is performed by merging data points of independently acquired two-dimensional data sets.

The application of ERI in permafrost-related studies is based on the high resistivity contrast between frozen and unfrozen conditions due to the weak electrolytic propagation of electric current in frozen materials (Kneisel et al., 2008). However, the phase transition of water is not only determined by temperature but also affected by soil properties like the salinity of the pore water or by interactions between the mineral matrix and water. Hence, the position of the freezing front corresponds not necessarily with the position of the 0 °C isotherm, which defines the occurrence of permafrost (Krautblatter et al., 2010; Scott et al., 1990).

For interpreting the resulting resistivity models, qualitative assessments of the resistivity-permafrost-relationship are used (e.g., Haeberli and Vonder Mühll, 1996; Ikeda and Matsuoka, 2006; Kneisel and Hauck, 2008). These assessments show that relatively high resistivity values indicate ice-rich permafrost and relatively low resistivity values indicate ice-poor permafrost. They further permit to distinguish between sedimentary ice and congelation ice.

Basic Theory

The subsurface is considered a porous material, where the electrical conductivity, and thus the electrical resistivity ρ as its reciprocal, is determined by (i) the resistivity of the host material, (ii) its porosity and (iii) the resistivity of the pore filling. This relationship is expressed by Archie's Law (eq. 3.1, Archie (1942)), which was initially applied for oil exploration purposes.

$$\rho_0 = \rho_f \phi^{-m} \tag{3.1}$$

Where ρ_0 is the resistivity of a fully water-saturated sample, ρ_f is the resistivity of the pore filling, ϕ the porosity fraction and m an empirically determined cementation exponent. However, a slightly adjusted form of Archie's Law (eq. 3.2), based on investigations by Winsauer et al. (1952), is more popular today, as studies showed that results of this equation align

Table 3.1: Resistivity Values for Different Subsurface Materials. Data from Kneisel and Hauck (2008), Reynolds (2011), Telford et al. (1990).

Material	Range of Resistivity [kΩm]
Sand	0.1–5
Gravel	0.1–1.4
Granite	5–1000
Mica Shist	0.2–10
Frozen Sediments, Ground Ice, Mountain Permafrost	1–1000
Glacier Ice (Temperate)	2000–1200000
Basalt	0.01–13000

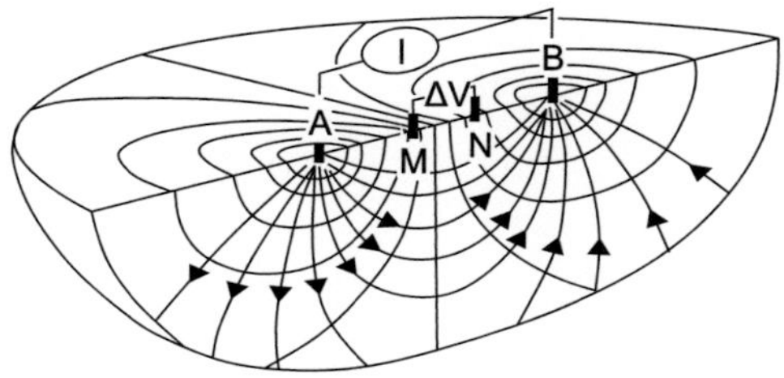

Figure 3.1: Current Flow in a Homogenous Halfspace. Modified from Lowrie (2011).

better with experimental data (Glover, 2016). This version of Archie's Law introduces the additional, empirically determined parameter a so it is stated:

$$\rho_0 = a\rho_f\phi^{-m} \tag{3.2}$$

Due to the omnipresence of water in most environmental settings and the relatively poor conductivity of mineral grains, the most important way of current flow in the subsurface is by electrolytic conduction. It occurs in the pore space, where water-solved ions serve as carriers of current transport. Alternative ways of conduction like (i) electronic conduction (i.e., current propagation through metal conductors), or (ii) dielectric conduction (i.e., current propagation through non-conductors under a changing electric field), are of minor importance in practice (Telford et al., 1990).

These considerations show a close relationship between resistivity and the actual pore filling of a material, particularly whether it is in solid, gaseous or liquid state. Variations of the pore filling therefore explain the wide range of characteristic resistivity values of different earth materials (tab. 3.1). However, this parameter, the so-called specific resistivity, cannot be measured directly from the ground surface. To conclude on this parameter, values of the apparent resistivity are measured. It must be noted here that the apparent resistivity is not an actual physical parameter of the subsurface, and that it depends on a mixture of different values of specific resistivity in an inhomogeneous subsurface (Lowrie, 2007).

To measure the apparent resistivity ρ_a, an electric direct current, generated with an artificial power source, is injected into the ground via two current electrodes (A, B). This results in the formation of an electric field within the subsurface (fig. 3.1). Then, electric potentials are measured at two other electrodes (M, N) with a resistivity meter. As the intensity (I)

Table 3.2: Advantages and Disadvantages of Different Electrode Arrays. Modified from Binley (2015).

	Dipol-Dipol (Dip-Dip)	Wenner-Schlumberger (WenSl)
Description	Current and potential electrodes arranged in separate pairs on a horizontal line.	Current electrodes enclose the potential electrodes on a horizontal line.
	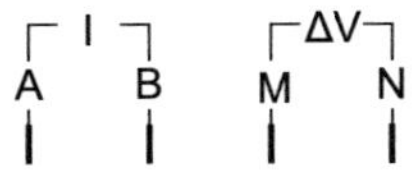	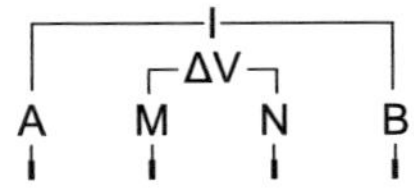
Survey Procedure	The separation between the two electrode pairs is increased.	The separation between the two current electrodes is increased.
Penetration Depth	Low	Medium
Lateral Resolution	High	Medium
Signal Amplitude	Low	Medium
Survey Speed	High (capable of multi-channel optimization)	Low

of the injected current is known and the difference of the electric potentials (ΔV) between the potential electrodes can be calculated from the measurements, it is possible to calculate a value for the apparent resistivity through Ohm's law (eq. 3.3).

$$\rho_a = (\frac{\Delta V}{I})k \tag{3.3}$$

Where the parameter k summarizes information on the electrode geometry and depends on the distance between the electrodes A and M (r_1), the distance between the electrodes B and N (r_2), the distance between the electrodes A and N (r_3) and the distance between the electrodes B and N (r_4) (see eq. 3.4) (Reynolds, 2011).

$$k = \frac{2\pi}{1/r_1 - 1/r_2 - 1/r_3 + 1/r_4} \tag{3.4}$$

Various configurations for possible arrangements of the four electrodes exist, the so-called arrays. As the sensitivity pattern of the different arrays varies, each array has specific advantages and disadvantages (see tab. 3.2).

Data Acquisition

The usage of multi-core cables allows to use dozens or hundreds of electrodes in one ERI survey. For the investigations presented in this thesis, either 36 or 72 electrodes were used. Following a predefined sequence, the resistivity meter selects two electrodes as current electrodes and two electrodes as potential electrodes. This combination of electrodes is then used to measure one data point of apparent resistivity. To ensure data quality, each data point is measured at least a second time with reversed polarization, and the deviation between the reciprocal measurements is stored. After one data point is measured, the resistivity meter switches to another combination of electrodes.

Multi-channel resistivity meters enable the simultaneous measurement of multiple datum points. While current is injected through one pair of current electrodes, electric potentials can be measured instantly at multiple pairs of potential electrodes. This option strongly

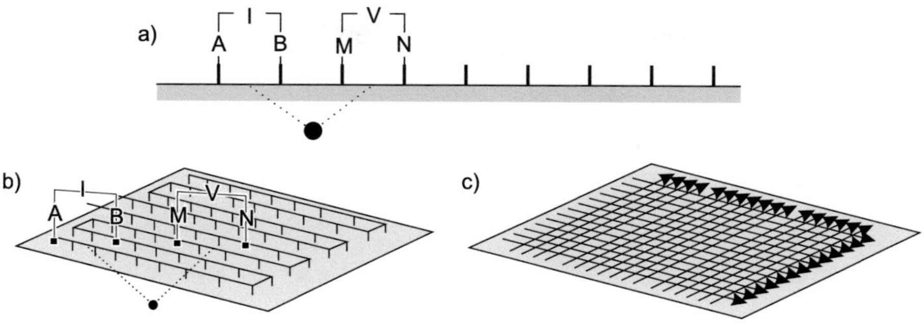

Figure 3.2: Different Types of ERI Data Acquisition. a) 2-D by Linearly Aligned Electrodes, b) r-3-D by a Rectangular Field of Electrodes, c) q-3-D by a Network of 2-D Lines (see Panel a). Own sketch.

increases the speed of data acquisition, but is bound to the Dipol-Dipol array, as the required enclosing of the potential electrodes in the Wenner-Schlumberger array rules out this optimization method (Dahlin and Zhou, 2004). Furthermore, only one of the two resistivity meters that were used for surveying is capable of multi-channel opitimzation: while the Syscal Pro resistivity meter (Iris Instruments S.A.S.) supports this option, it is not supported by the Syscal Junior Resistivity Meter (Iris Instruments S.A.S.) for technical reasons. To ensure a sufficient ground coupling between the electrodes and the subsurface, which enables the acquisition of reliable data, ground resistance values are checked at every electrode before the measurements are started. At electrode positions where exceptionally high resistance values (e.g., $>30\,\mathrm{k\Omega m}$) are measured, water-soaked sponges are attached to the electrodes (Marescot et al., 2003). However, this procedure is not recommended at ERTM locations where measurements with permanently installed electrodes are performed, to avoid any external influence on the data.

Two basic types of data acquisition can be distinguished, (i) two-dimensional data acquisition and (ii) three-dimensional data acquisition. For two-dimensional data acquisition, the electrodes are linearly aligned (fig. 3.2a). Data points of apparent resistivity are hence arranged in a two-dimensional profile. For three-dimensional type of data acquisition, the electrodes are arranged in a horizontal field, and the resulting data points are hence arranged in a three-dimensional space (fig. 3.2b). This type of data acquisition is usually restricted to small areas, as the number of electrodes that is required for a meaningful data coverage is increased to the square and the required cable length is also strongly increased. However, this setup enables the acquisition of data points from diagonal or oblique alignments of electrodes, which is beneficial for the accuracy of the resulting resistivity model (Loke, 2016b). Although it is possible to set electrodes at arbitrary positions, constant separations between the electrodes are commonly used.

An approach that permits the generation of spatially extensive three-dimensional resistivity models, is to merge data from a network of independently performed parallel and perpendicular two-dimensional surveys (fig. 3.2c) (see e.g., Bentley and Gharibi, 2004; Chambers et al., 2002). The term q-3-d, which is used for this approach in this thesis, refers to the two-dimensional type of data acquisition, but emphasizes that the data set is subsequently treated as one three-dimensional data set. A crucial factor to avoid undersampling in the

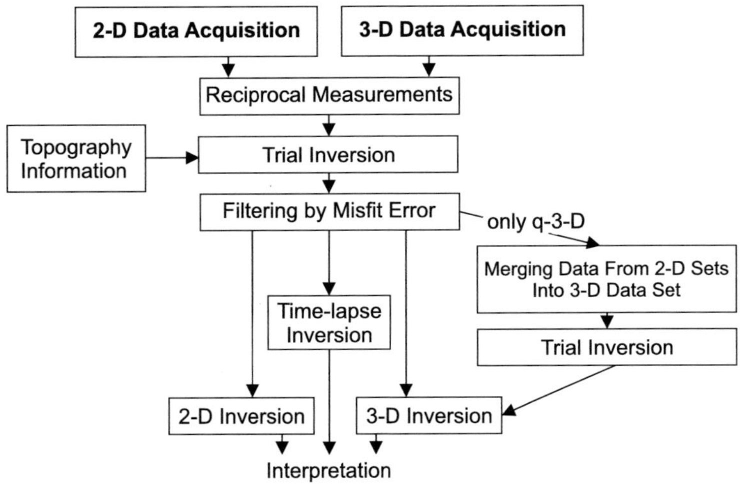

Figure 3.3: Workflow for Processing ERI Data. Own Sketch.

areas between the two-dimensional survey lines is their separation. This line-separation is recommended not to exceed twice the electrode spacing (Gharibi and Bentley, 2005).

The models presented in this thesis were gained by following this recommendation in one survey direction. In the perpendicular direction, adjustments to line separation were necessary in certain situations, e.g., when deep snow fields or huge boulders prevented the setup of a survey line at the recommended distance. However, as the setup of perpendicular lines is not compulsory for three-dimensional approaches (Loke et al., 2013), it is assumed that the acquired data sets provide a sufficient data coverage.

Data Processing and Inversion

The basic workflow for processing the acquired data is displayed in figure 3.3. As stated in the previous section, the acquired data sets are quality checked through reciprocal measurements during data acquisition. Only data points with a deviation below 5 % between these reciprocal measurements are used for further processing. Information on surface topography, gained either by Real-Time Kinematic (RTK) positioning data from a GNSS (Global Navigation Satellite System) or estimated visually in the field, is subsequently added to the resistivity data sets. Following the procedures suggested by Loke (2016b) a trial inversion is carried out on each data set and outlying data points, as which all data points that produce a misfit error $\geq$100 % between the measured and the modeled resistivity values are considered, are eliminated. While the filtered data sets of 2-D and r-3-D data acquisition approaches are inverted directly after this step, the two-dimensional data sets used for q-3-D ERI have to be merged and another trial inversion is performed on the resulting three-dimensional data sets. Processing of ERTM data sets is similar to processing of 2-D data sets.

Modeling the distribution of specific resistivity from a set of measured apparent resistivity values is a non-linear and ill-conditioned problem (Pidlisecky et al., 2007). To solve this problem, inversion schemes are used. These techniques, which require an initial model, are implemented in different inversion software products. Multiple products are available, licensed both proprietary and open source. The first group includes software products like EarthImager3D (Advanced Geosciences, Inc.), RES2DINV/RES3DINV (Geotomo Software Sdn. Bhd.) or ZondRes3D (Advanced Geophysical Operations and Services, Inc.). In contrast to these commercial products, software like e.g., BERT (Boundless Electrical Resistivity Tomography) (Günther et al., 2006; Rücker et al., 2006) or RESINVM3D (Pidlisecky et al., 2007) are free and open source. All software products execute two main tasks: (i) forward modeling and (ii) inverse modeling.

Forward modeling is used to calculate values of apparent resistivity from a model of specific resistivity values. This step is required for the inversion scheme, which calculates the misfit between the observed and the modeled data. Parameters that are required for an accurate forward modeling of resistivity data are (i) topography, (ii) resistivity of the subsurface and (iii) information on the electrode arrays. The modeling routine divides the subsurface into cells and assigns values of electrical potential to them. For this purpose, different methods are used, commonly either the (i) finite-difference or the (ii) finite-elements method. While the first method is based on differential approximations to the partial derivatives, the second method is based on integral approximations (Binley, 2015). Main advantage of the more complex finite-element approach is the possible implementation of arbitrary model geometries which allows a more accurate incorporation of topographical data.

The solution of the inverse-problem, also referred to as inverse modeling, describes the generation of a geophysical model of which the measured data can be reproduced. This relationship is underdetermined, as the number of measured data points of apparent resistivity is finite, but the subsurface distribution of specific resistivity is considered continuous. Consequently, many models can satisfactorily reproduce to the observed data, especially as the measured data is affected by noise (Pidlisecky et al., 2007). To overcome this problem, common inversion schemes use a least square approach. A basic formulation of this approach to minimize the data-model misfit ϕ_d is given in equation 3.5 (Binley, 2015), where d is the measured data and W_d is a diagonal matrix with entries equal to the standard deviation of the measurements, which is used for data weighting:

$$\phi_d = (d - d_{pred})^T W_d^T W_d (d - d_{pred}) \tag{3.5}$$

To ensure stability of the inverse modeling, a regularization parameter λ is used. It is required to avoid that small variations in the measured data lead to large changes in the final model (Li and Oldenburg, 1999). The regularization parameter is a tool that can be adjusted according to prior knowledge and needs to be chosen with respect to data quality (Cockett et al., 2015). To describe the effect of regularization, the model roughness parameter ϕ_m, calculated from a roughness matrix R and the model parameter m, is introduced and included into the formula of total data misfit (eq. 3.6) (Binley, 2015):

$$\phi_{total} = \phi_d + \lambda \phi_m \tag{3.6}$$

While a high regularization leads to smooth models with high misfit errors, a small regularization results in a heterogeneous pattern of resistivity values. In common software products, the user can (i) determine regularization manually, or (ii) use an automatic approach to determine regularization. A comparison between different automatic approaches is presented by

Farquharson and Oldenburg (2004). Most software products use iterative approaches based on a Gauss-Newton scheme for minimizing the data misfit. This procedure stepwise assigns values to the model cells in order to minimize the difference between the calculated and the observed resistivity values (eqs. 3.7 and 3.8) (Binley, 2015):

$$(J^T W_d^T W_d)\lambda m = J^T W_d^T (d - d_{pred_k}) - \lambda R m_k \tag{3.7}$$

$$m_{k+1} = m_k + \Delta m \tag{3.8}$$

where J is the sensitivity or Jacobian matrix with the components J_{ij}, that states the change of the ith datum point with respect to the jth model parameter. The parameter set at iteration k is represented by m_k and Δm is the parameter change at iteration k. For minimizing the differences between the modeled and the observed values, two methods are commonly used: (i) the L1-norm or robust inversion, or (ii) the L2-norm or smoothness-constrained least square inversion. While the first method tries to minimize the sum of the absolute values of the data misfit, the second method minimizes the sum-of-squares. In addition to equation 3.7 and equation 3.8, which describe an L2-norm inversion, the L1-norm inversion uses iteratively reweighted matrices for the observed and the modeled data to reduce the effect of bad data points and to produce models with sharp boundaries (Loke et al., 2003).

The inversion procedure continues until it reaches a convergence criterion. This criterion is an indication that the model cannot be improved any further. Common software products offer different criteria for terminating the inversion, as the assumption that the model has reached an ideal level depends on hardly appraisable parameters like noise (Pidlisecky et al., 2007). Commonly, the inversion is stopped (i) after a fixed number of iteration steps or (ii) when a calculated value, chosen with respect to data misfit and error bounds, is reached. Beside mathematical measures, it must be stated that the model with the lowest misfit error is not necessarily the best model from a geological perspective (Kneisel et al., 2008).

To evaluate if a resistivity model is sufficiently resolved, the application of a resolution matrix approach (Emmert and Kneisel, 2017; Loke, 2016b; Stummer et al., 2002; Wilkinson et al., 2006) is recommended. This approach estimates the information content of each model cell and provides a measure on the influence of inversion settings and neighboring model cells. In the most popular software package RES2DINV/RES3DINV, resolution matrix values are transferred into index values, taking into account the discretization of the model (Loke, 2016a). The resistivity models presented in this thesis were processed with the software products RES2DINVx64 (Ver. 4.05.32) and RES3DINVx64 (Ver. 3.11.57). For the comparative analysis between two inversion software products (see chap. 10), BERT (Ver. 2.1.1) was used.

Comparison of Inversion Software

This thesis includes a comparison between resistivity models from two different software: (i) the wide-spread commercial RES2DINV/RES3DINV software package and (ii) the BERT inversion software, published under GNU Public License 3 and free for academic purposes. However, a comparison is difficult as the code of RES2DINV/RES3DINV is not published. Key parameters which differ between the two software products include (i) discretization (ii) regularization, and (iii) the convergence criterion.

- RES2DINV/RES3DINV subdivides the subsurface into cuboid-shaped cells that reflect the electrode spacing of the survey setup. Different options allow to refine the mesh in horizontal dimension by increasing the number of nodes between the grid lines and in vertical dimension by adjusting the number of model layers. BERT, in contrast, uses irregularly shaped tetrahedral cells. This approach considers the actual model resolution and enables a model refinement in areas of high sensitivity. Furthermore, it saves computation time, as a global refinement is unnecessary. For these reasons, the approach that is used by BERT is considered to be superior to conventional approaches that use regularly shaped model cells (Binley, 2015).
- The provided options to adjust the regularization setting differ between the two software products. When a manual adjustment is used, the user of RES2D/RES3DINV can choose a starting value for the damping factor which is decreased during the inversion process until a pre-determined minimum value is reached. Additionally, a higher damping factor can be used for the uppermost model layer. By default, BERT uses a constant regularization value, but there is also a similar option to use a pre-defined damping factor that decreases during the inversion process. Both software products provide additional options for an automatic estimation of the regularization strength. For this task, BERT uses an L-curve approach, in which the software estimates an ideal damping factor from plotting values of data misfit for different damping factors against model roughness (Günther et al., 2006). The optimization approach used by RES3DINV/RES3DINV is not published, but according to Loke (2017, personal communication), it is based on a similar approach.
- In RES2DINV/RES3DINV, the inversion process is terminated when the change in the overall misfit error between two iteration steps is lower than a pre-defined value. BERT also stops the inversion process at a level of stagnating minimization, but uses a different criterion based on a χ^2-method (Günther et al., 2006). However, both software products offer the option to manually define a maximum number of iterations that is carried out.

For the comparative analysis, synthetic geocryological models of two landform types, one rock glacier (SYN01, fig. 3.4a) and one palsa area (SYN02, fig. 3.4b) were created with the toolkit provided by the software RES3DMOD (Geotomo Software Sdn. Bhd., Version 2.14.23 Plus). This toolkit allows the creation multi-layered 3-D subsurface models with a high number of cells to which the user can assign individual resistivity values, based on literature values or user experience. Synthetic modeling is commonly performed as part of the back-and-forth procedure, presented e.g., by Hilbich et al. (2009), in which synthetic models are used to constrain the inversion of measured data. It is further used in studies that target methodological questions, like e.g., evaluating the imaging capabilities of different electrode arrays (Dahlin and Zhou, 2004). Since RES3MOD is not able to integrate topographic data into the synthetic resistivity models, both models have a flat surface topography.

The calculation of apparent resistivity values from the synthetic models was based on networks of linearly aligned electrodes, which corresponds to data acquisition for a q-3-D ERI data set. A noise level of 5 % was added to each data set. The data files were subsequently adjusted individually to meet the input requirements of the two software products. Similar

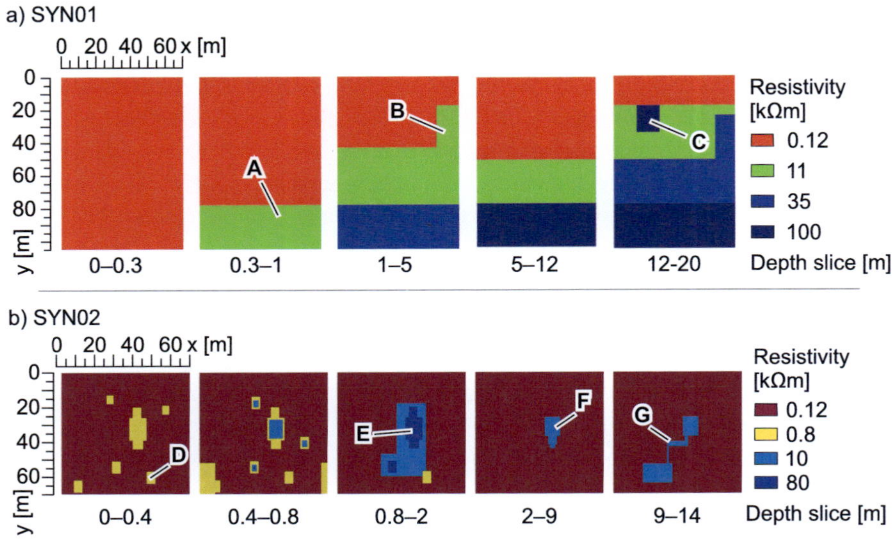

Figure 3.4: Synthetic Resistivity Models. a) SYN01, b) SYN02.

inversion settings and procedures were used to approach a high level of comparability: a relatively low level of regularization and an L1-norm inversion were chosen. Although the models represent a flat surface topography, finite-element modeling was used in the step of forward modeling. The range of modeled resistivity values was not limited, and the inversion process was stopped manually after five iteration steps in both software products.

Resistivity values of the synthetic rock glacier model (SYN01) are between 1.2 kΩm and 100 kΩm (fig. 3.4a). One structure of high resistivity values covers one side of the model (A). Resistivity values increase with depth and reach a maximum below a depth of 5 m. The extent of structure A increases at depths between 1 m and 5 m, and a small, longitudinal band of high resistivity values (B) attaches to structure A. Structure B is absent between 5 m and 12 m, but it reappears at depths between 12 m and 20 m. At the same depth level, an isolated patch of cells with resistivity values of 100 kΩm (C) appears.

Resistivity values of the synthetic palsa model SYN02 are between 0.1 kΩm and 80 kΩm (fig. 3.4b). While multiple patches of relatively high resistivity values are present in the shallow subsurface (D), resistivity values of up to 80 kΩm cluster at one spot between depths of 0.8 m and 2 m (E). Between depths of 2 m and 9 m, the extent of structure E reduces, and resistivity values of the model cells are not exceeding 10 kΩm (F). In the lowermost model layer, cells with high resistivity values cover again a larger area, with a complex outline (G).

3.2 Ground-Penetrating Radar

In addition to electric, seismic and electromagnetic methods, Ground-Penetrating Radar is one of the four main geophysical methods (Hauck, 2013). It is based on the transmission of artificially created electromagnetic (EM) pulses into the ground, which are reflected at

Table 3.3: Specific Dielectric Constants and EM Velocity Values for Different Materials. Data from Davis and Annan (1989); Reynolds (2011).

Material	Dielectric Constant	EM Velocity [$\mathrm{m\,ns^{-1}}$]
Air	1	0.3
Fresh water	80	0.033
Dry sand	3–5	0.15
Saturated sand	20-30	0.06
Limestone	4–8	0.12
Shales	5–15	0.09
Ice	3–4	0.16

layer boundaries in the subsurface. The echoes of the reflected waves, the so-called Two-way Traveltimes (TWT) are recorded and transferred into an image of the subsurface layering (Scott et al., 1990). In this thesis, GPR is used as a complementary method to ERI, as results of GPR provide supplementary information on the subsurface layering.

The application of GPR in permafrost-related investigations is based on its ability to differentiate between frozen and unfrozen subsurface conditions by the detection of thermal boundaries, which are independent from changes in the material composition. This is possible as the physical properties of liquid water and ice differ strongly (Moorman et al., 2003). GPR therefore permits a differentiation between massive ice and ice-saturated conditions as well as between different ways of material deposition (Scott et al., 1990). For such questions, characteristic reflection patterns, so-called "radar facies" are used (Van Overmeeren, 1998).

Basic Theory

Differences in the electromagnetic (EM) properties between different subsurface materials affect the dynamics of propagating EM waves. A crucial variable for the successful application of GPR is the EM velocity v of the propagating waves. Its value can be calculated from the speed of light c and the dielectric constant (or relative electrical permittivity) ϵ_r of the material that is crossed by the EM waves (Everett, 2013) (eq. 3.9).

$$v = \frac{c}{\sqrt{\epsilon_r}} \tag{3.9}$$

The dielectric constant depends primarily on the water content and the level of water saturation of the specific materials, but also on the physical state of the water. Exemplary values of dielectric constants and EM velocities are stated in table 3.3. When artificially created EM waves are reflected at a boundary between two layers that have the dielectric constants ϵ_{r1} and ϵ_{r2}, respectively, the reflection coefficient R is calculated from the ratio between the dielectric constants, as stated in equation 3.10 (Berthling and Melvold, 2008). To enable a detection of these reflections, Annan and Cosway (1992) claim that R_2 should be at least 1/100 and that the ratio between the depth of the second layer and its smallest lateral dimension should not exceed 10/1.

Beside a reflection of the EM waves, their amplitude A diminishes when they travel through a subsurface medium. The strength of this signal loss depends on the travel distance

z and the attenuation constant α (see eq. 3.11), which, in turn, depends on the electrical conductivity σ, the magnetic permeability μ and the dielectric constant ϵ (see eq. 3.12) (Berthling and Melvold, 2008).

$$R \approx \frac{\sqrt{\epsilon_{r1}} - \sqrt{\epsilon_{r1}}}{\sqrt{\epsilon_{r1}} + \sqrt{\epsilon_{r1}}} \tag{3.10}$$

$$A = A_0 e^{-\alpha z} \tag{3.11}$$

$$\alpha = \frac{\sigma}{2}\sqrt{\frac{\mu}{\epsilon}} \tag{3.12}$$

In addition to a sufficiently high reflection coefficient R, the ability to detect discrete structures in the subsurface is bound to the size of the target, which must exceed a certain radius r. This radius depends on the depth of the structure h below the surface and the wavelength λ of the EM waves (see eq. 3.13) (Annan, 2009). The wavelength, in turn, depends on the EM velocity of the surrounding materials and the frequency of the transmitted EM pulses. Berthling and Melvold (2008) present an example for the estimation of a minimum reflector radius, given a frequency of 50 MHz and a EM velocity of $0.15\,\mathrm{m\,ns^{-1}}$. It results in a wavelength of 3 m and a minimum radius of 5.5 m for a reflector at a depth of 20 m.

$$r = \sqrt{\frac{\lambda h}{2} + \frac{\lambda^2}{16}} \tag{3.13}$$

Data Acquisition

During GPR data acquisition, EM pulses are transmitted into the subsurface with a constant frequency by a transmitter antenna (Tx). A receiver antenna (Rx) detects incoming echoes of the EM waves and records the amplitude and the propagation time of the reflected signals (see fig. 3.5). Two different modes of GPR surveying are used for the investigations in this thesis: (i) one-dimensional Common Mid-Point surveying (CMP) and (ii) two-dimensional Common Offset profiling (abbreviated as *2-D GPR* in the following).

In CMP mode (fig. 3.5a), transmitter and receiver antenna are moved away from a Common Mid-Point with a fixed step size. This enables an estimation of the vertical EM velocity changes in the subsurface and thereby a characterization of the subsurface layering (Van der Kruk, 2015). Disadvantages of this approach are (i) that it provides information only in one-dimension, and (ii) that an undisturbed subsurface layering is required through the complete area of investigation (Reynolds, 2011). In periglacial research, CMP surveying is used to create one-dimensional models of the vertical EM velocity layering for the conversion of recorded time values into depth values.

In Common Offset mode (fig. 3.5b), transmitter and receiver antenna are separated by a fixed distance and the entire setup is moved along a horizontal line on the ground surface with a constant step size. The resulting two-dimensional cross-section of the subsurface enables the detection of discrete structures and their horizontal and vertical extent (Van der Kruk,

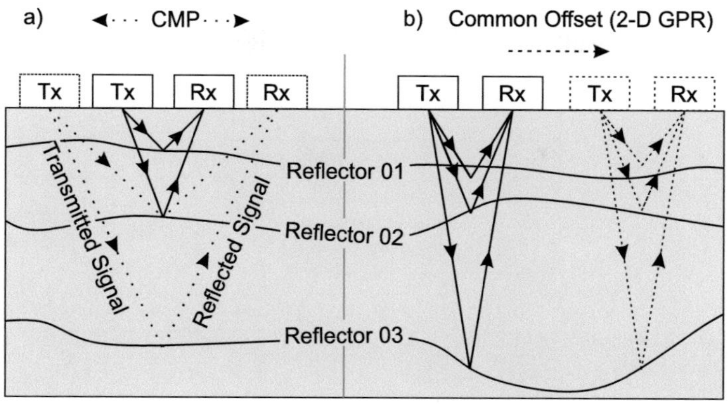

Figure 3.5: Different Types of GPR Data Acquisition. a) Common Offset, b) Common Mid-Point. Modified from Reynolds (2011).

2015). Three-dimensional data sets can be created by merging data from multiple closely separated, parallel Common Offset surveys. The separation between the parallel survey lines must be chosen with respect to the extent of the survey area and the targeted level of detail (Bristow, 2009). To achieve a so-called "full-resolution 3D GPR Imaging" (Grasmueck et al., 2005), however, the line separation must be smaller than $1/4\lambda$, which results in values between 0.1 m and 0.2 m, when 100 MHz antennas are used (Grasmueck and Weger, 2002; Grasmueck et al., 2004).

A Pulse EKKO Pro System (Sensors&Software Inc.), connected to antennas with center frequencies of 50 MHz, 100 MHz, or 200 MHz, was used for data acquisition in this thesis. When an EM velocity of $0.12\,\mathrm{m\,ns^{-1}}$ is assumed, these frequencies enable a detection of structures that exceed a vertical extension of 0.6 m (50 MHz) 0.3 m (100 MHz) and 0.15 m (200 MHz), respectively (Bristow, 2009). For all profiling surveys, antennas were orientated parallel to each other and perpendicular to the survey direction, in the so-called perpendicular broadside orientation (Barker and Moore, 1998).

Data Processing

While processing of 1-D CMP data requires only a time-zero correction (see following paragraph) for reliable estimations of EM velocity values (Sandmeier, 2017), GPR profiling data requires a more sophisticated processing. The workflow presented in fig. 3.6 was used for processing the data sets presented in this thesis. The steps were carried out with the commercial software ReflexW (Sandmeier geophysical research, Version 8.5.3).

After data acquisition, a first processing step is the elimination of duplicated traces, which can occur e.g., from accidentally pressing the start button twice (Berthling and Melvold, 2008). A second step is to shift the recorded signal traces along the time axis, so that the first recorded EM pulse is set to a certain time value, commonly to zero. This procedure removes temporal displacements of single traces, which can be caused e.g., by variations of the antenna orientation.

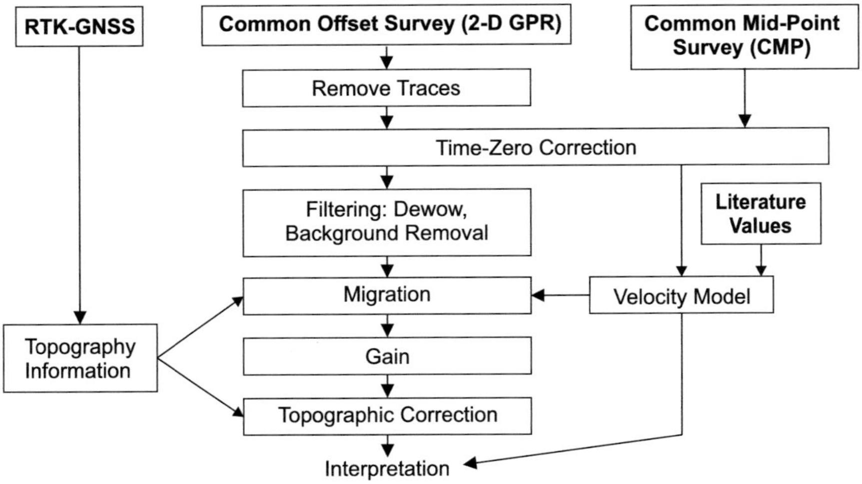

Figure 3.6: Workflow for Processing GPR Data. Modified from Berthling and Melvold (2008).

Subsequent filtering steps include the so-called *Dewow*-filtering, which removes unwanted components of very low frequency caused by electromagnetic induction, and the removal of a moving average of the signal amplitude, the so-called *background-removal*. This filtering step removes signals caused by a direct coupling between the antennas and contain no information on the subsurface (Everett, 2013).

After filtering the data, *migration* compensates distortion effects on wave propagation in the subsurface, caused by point scatterer or dipping features. This processing step relocates reflections from their measured positions on a single trace to their actual positions, which are calculated with respect to neighboring traces (Van der Kruk, 2015). Among the available migration techniques, the *topographic migration*, works with respect to a distinct surface topography and is hence preferred for data from high-alpine sites (Sandmeier, 2017).

Following to the step of migration, the recorded TWT values are transferred into values of depth (*time-depth conversion*). This step requires information on the vertical EM velocity layering in the subsurface (Berthling and Melvold, 2008), which can be derived from (i) CMP surveying, (ii) hyperbola matching or (iii) literature values. For the second option, the shape of reflection or diffraction hyperbolas that appear in Common Offset radargrams when EM waves are reflected by single objects, is investigated. This allows to assess the EM velocity in the surrounding of the scatterer (Cassidy, 2009). If hyperbolas are absent in the radargram, estimations on the vertical EM velocity layering that are based on literature values can be used to produce reliable results (Cassidy, 2009).

To compensate a strong signal attenuation, especially of reflections from deeper parts of the subsurface, the application of *gains* is advised. This improves the visibility of weak reflections, but as it can change the relative reflection amplitude and thereby alter the structure of the data, gains must be used carefully (Cassidy, 2009). Finally, the radargrams are displayed with topography. Merging data from multiple 2-D GPR surveys into one three-dimensional data set is also enabled by REFLEXW. The workflow for processing three-dimensional data sets is comparable to the presented workflow for two-dimensional data sets.

3.3 Additional Methods

In addition to the geophysical data, this thesis comprises records of subsurface temperature values, measured as vertical temperature profiles within boreholes, surface temperature values and values of the soil matric potential.

Temperature Measurements

The continuous recording of subsurface temperature values provides a direct proof of permafrost, according to definition stated at the beginning of this thesis. Nevertheless, records of Ground Surface Temperature (GST) values can also provide information on the subsurface conditions. This is because the ground surface is the interface between the atmosphere and the subsurface and thereby controls the intensity of energy exchange processes (Haeberli, 1975; Haeberli and Patzelt, 1982).

Temperature values at the ground surface were recorded hourly by M-Log data loggers equipped with *PT1000* sensors that provide an accuracy of ±0.1 °C (GeoPrecision GmbH). Data loggers were distributed at topographically different surface positions in the investigation areas and placed (i) either at depths of a few centimeters or (ii) below small stones to shield them from solar radiation. The same type of data logger was used to record subsurface temperature values, which were measured hourly by thermistor chains equipped with 15 *Dallas* sensors (Geoprecision GmbH). The vertical positions of these sensors, which provide an accuracy of ±0.25 °C, are the ground surface and depths of 0.2 m, 0.4 m, 0.8 m, 1.2 m, 1.6 m, 2 m, 3 m, 4 m, 5 m, 6 m, 7 m, 8 m, 9 m and 10 m.

Values of the mean daily and the mean annual ground surface temperature (MAGST) were calculated from the recorded data. For assumptions on the existence of an insulating snow cover, diurnal temperature variations were analyzed (Lewkowicz, 2008; Reusser and Zehe, 2011). When diurnal temperature fluctuations are below 0.4 K throughout one complete day, the existence of an insulating snow cover is assumed (Rödder and Kneisel, 2012a). So-called *zero-curtain* periods describe time spans in which surface temperatures are nearly constantly around 0 °C. Their appearance is caused by latent heat that is bound in the phase change of water in times of freezing or thawing (French, 2018; Rödder and Kneisel, 2012a).

Soil Matric Potential Measurements

The amount of liquid water in the subsurface is an important variable that characterizes the ground thermal regime in frozen areas (Hauck et al., 2008; Kane et al., 2001). To approach this variable in the presented investigations, measurements of the soil matric potential with *Tensiomark* sensors (EcoTech Umwelt-Messsysteme GmbH) were used. Data was acquired hourly and recorded by a data logger. This approach is relatively innovative in the field of periglacial geomorphology. Compared to approaches where the volumetric water content is measured (Kujala et al., 2008; Pellet et al., 2016; Rist and Phillips, 2005), the approach of measuring the soil matric potential is not bound to small sample volumes, which may be unrepresentative in case of heterogeneous settings. In contrast to measurements with tensiometers (e.g., Matsuoka, 1996), the device is the frost resistant.

The measurement principle is based on moisture-related variations in the heat capacity of a porous ceramic plate, which is measured with an artificial heat pulse. As this requires a simultaneous recording of the temperature in the surrounding of the sensor, the sensor provides additional information on subsurface temperatures (ecoTech, 2014). The resulting pF-values indicate the level of energy that is required to pull water out of the subsurface that is held by capillary and absorptive forces. A pF-value of 0 represents a state of full water saturation while a pF-value of 7 represents extremely dry conditions (Blume et al., 2016).

3.4 Surveys Maps and Setups

The last section of this chapter presents individual maps of the six study sites, which show the locations of the geophysical surveys and the positions of the surface/subsurface temperature sensors and the matric potential sensors. Further details on survey specifications, like e.g., the electrode spacing that was used or the dates on which the surveys were performed, as well as details on the results of the geophysical modeling, are presented in appendix A. At the end of this section, the synthetic models for the comparative analysis between two software products are displayed.

Survey Setups NAR

Between 2014 and 2017, two q-3-D ERI, 18 2-D ERI, four 2-D ERTM and two 2-D GPR surveys were performed on the rock glacier assembly at NAR study site to investigate the distribution of permafrost and to assess the internal structure of the rock glaciers (fig. 3.7). Information on the vertical EM velocity layering, which is required for time-depth conversion of the GPR data, is gained by one 1-D CMP survey. Geoelectrical monitoring surveys were performed between August 2015 and October 2016 at NAR01 (E02/E03) and between July 2016 and October 2016 at NAR07 (E08/E09) (see tab. A.3, appendix). Subsurface temperature values were measured with one thermistor string, installed in a borehole with a depth of 10 m, and recorded between September 2014 and August 2017. Surface temperature data was measured by ten GST sensors, which were placed at different topographic positions. Except for the T07 sensor, which was installed only on October 15, 2015, GST data was recorded between September 2014 and August 2017. The sensors T05, T07 and T10 stopped recording in June 2016 due to battery failure, and due to the same reason, a small data gap exists at T04. In addition to geophysical surveying and temperature recording, core drilling was performed at one location on NAR06 in September 2016.

Survey Setups UER

At UER study site, three q-3-D ERI surveys were performed on UER01 in the years 2014 (E01), 2015 (E02) and 2016 (E03), respectively for a detailed investigation of the subsurface conditions. Four 2-D GPR surveys were additionally performed on UER01, one in longitudinal direction of the rock glacier (R01) and three in transverse direction (R02–R04). On UER02, only one 2-D ERI survey (E04) was performed in August 2015 (fig. 3.8). Temperature data at UER study site was recorded with a subsurface thermistor string installed in a borehole with a depth of 10 m at UER01, and at ten surface positions on UER01. Eight GST sensors were placed on the surface of UER01: they were distributed around the furrow-and-ridge topography of the rock glacier snout or placed at positions in the transitional area between the frontal and the central part of the rock glacier. Two additional GST sensors were placed on the apparently relict lobe in front of the rock glacier (T09) and on the scree slope in the east of UER01 (T10), respectively. Continuous temperature records are available throughout the time span from October 2014 to August 2017.

Survey Setups LTF

One q-3-D ERI survey, which covers both the coarse-grained and the fine-grained section of LTF rock glacier, was performed at LTF study site in 2014 (fig. 3.9). To provide complementary information, four 2-D GPR surveys were performed in 2016: one pair of parallel surveys was performed along the longitudinal axis of the rock glacier (R03, R04) and one pair of parallel surveys was performed in transverse direction (R05, R06), respectively. One 1-D CMP survey was performed in the area covered with fine-grained debris (R01) and one 1-D CMP survey was performed in the area where rather coarse-grained debris dominates the ground surface material composition (R02). These two 1-D CMP approaches provide information on the vertical EM velocity layering of the subsurface in both sections and were used to improve the reliability of the time-depth conversion.

Survey Setups FUR

To investigate subsurface conditions of the lobe-bearing talus slope, one q-3-D ERI survey was performed at FUR study site in 2014 (E01). For more detailed investigations, two smaller r-3-D ERI surveys were performed on the two selected lobes, FUR01 (E02) and FUR02 (E03), in 2015. In addition to the geoelectrical surveys, two 2-D Common Offset GPR surveys were performed in longitudinal direction on the upper (R01) and the lower (R02) lobe in 2017 (fig. 3.10).

Survey Setups OVR

At OVR study site, two r-3-D ERI, seven 2-D ERI, five 2-D GPR surveys and one 3-D GPR survey were performed in 2015 and 2017 (fig. 3.11). At three locations (T01–T03), GST sensors were placed for continuous temperature recordings. Their locations represent exposed positions in the surrounding desert (T01) and on a palsa (T02), respectively, as well as a rather sheltered position between two palsas in the central part of the wetland area (T03). At the locations of T02 and T03, additional Tensiomark sensors were installed at depths of 0.5 m (T02) and 0.55 m (T03), respectively. While GST values were recorded failure-free between September 2015 and September 2017, the Tensiomark sensor at T03 failed for an unknown reason in December 2016. In addition to geophysical surveying and the recording of temperature values and pF-values, the topography of the frost table was manually sampled within the extents of the surveys R01 and E01 with a 1.2 m long steel rod. This approach provides ground truth information that was used to assess the EM velocity of the subsurface materials for the GPR time-depth conversion.

Survey Setups HPA

One q-3-D ERI, one r-3-D ERI, one 2-D ERI and four 2-D Common Offset GPR survey were performed at HPA study site between 2015 and 2017 (fig. 3.12). Most of the surveys concentrate on HPA01, where sorted polygons occur. At HPA02, where surface cracks form polygonal structures in an accumulation of fine-grained materials, one r-3-D ERI and one 2-D GPR survey were performed. In addition to the geophysical surveys, GST data was continuously recorded at HPA01 between September 2015 and September 2017.

Survey Setups SYN

Virtual surveying on the synthetic rock glacier model SYN01 comprises 30 synthetic 2-D ERI surveys with Dipol-Dipol array (fig. 3.13a). The electrode spacing is 2 m for the lines in x-direction and 3 m for the lines in y-direction. This results in 12240 data points. As the separation between the parallel lines in x-direction, never exceeded the recommended value of twice the electrode spacing, the appearance of undersampling effects is presumably avoided. Virtual surveying on the synthetic palsa model SYN02 comprises 18 2-D ERI survey lines of the Wenner-Schlumberger array, which results in a total number of 5184 data points. In both horizontal dimensions, an electrode spacing of 2 m was used in the center of the investigated area. Towards the margins of the investigated area, the electrode spacing was increased to imitate a realistic survey setup. This setup is assumed to promote the occurrence of undersampling effects at the margins of the investigated area, as recommendations on line separation are ignored in these parts.

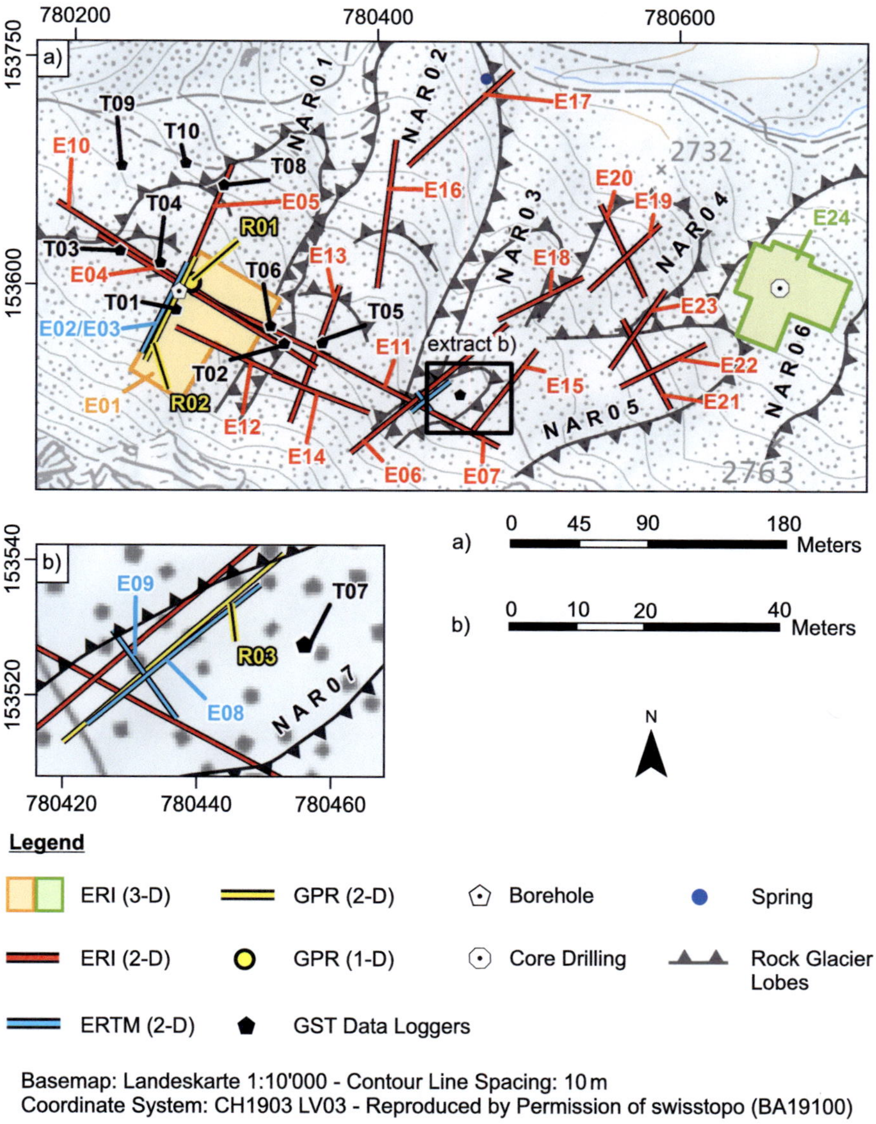

Figure 3.7: Map of Survey Locations NAR.

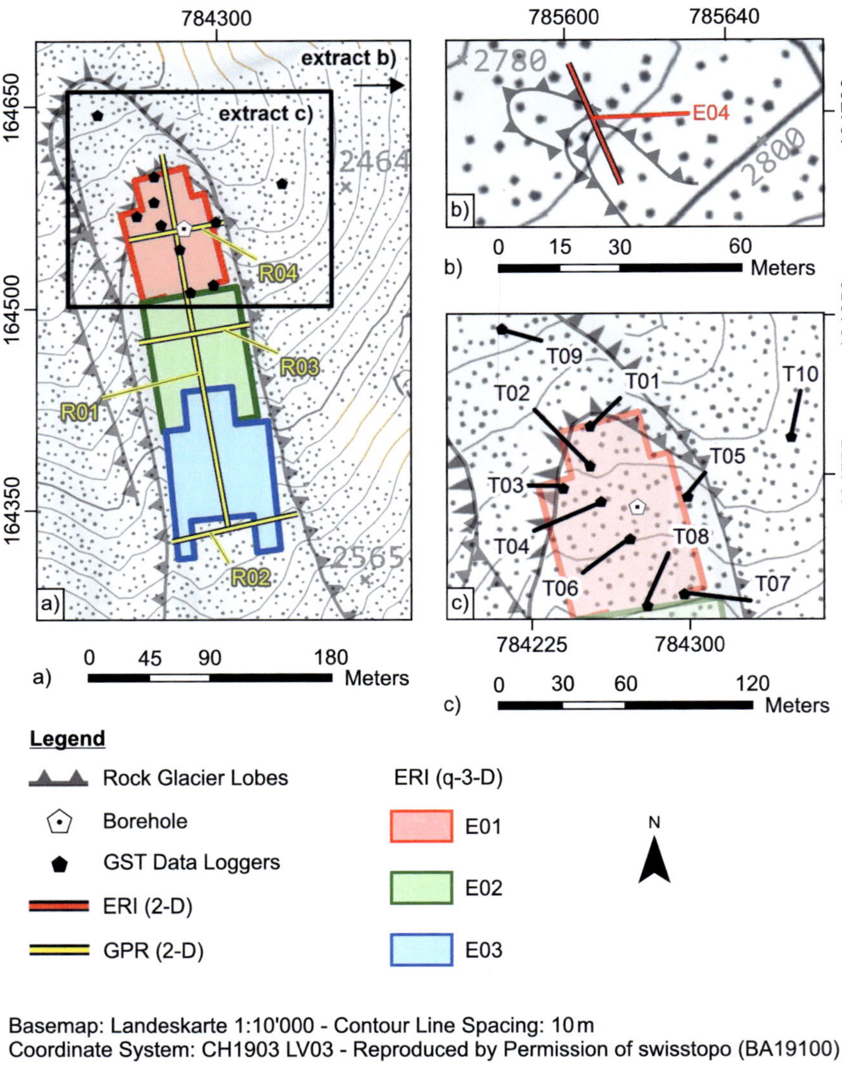

Figure 3.8: Map of Survey Locations UER.

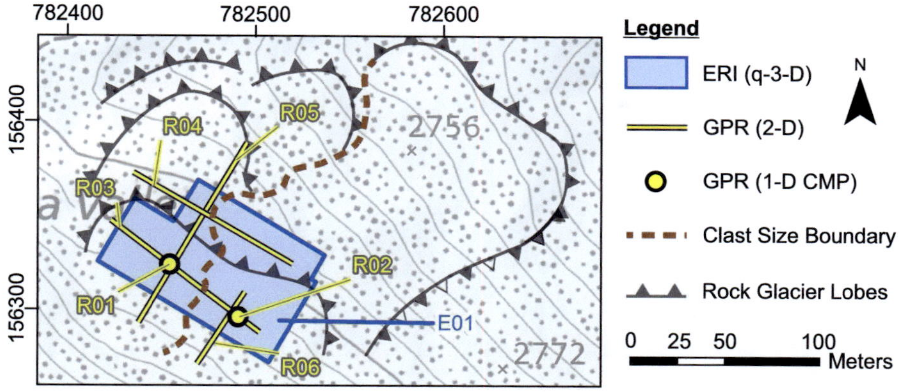

Basemap: Landeskarte 1:10'000 - Contour Line Spacing: 10 m
Coordinate System: CH1903 LV03 - Reproduced by Permission of swisstopo (BA19100)

Figure 3.9: Map of Survey Locations LTF.

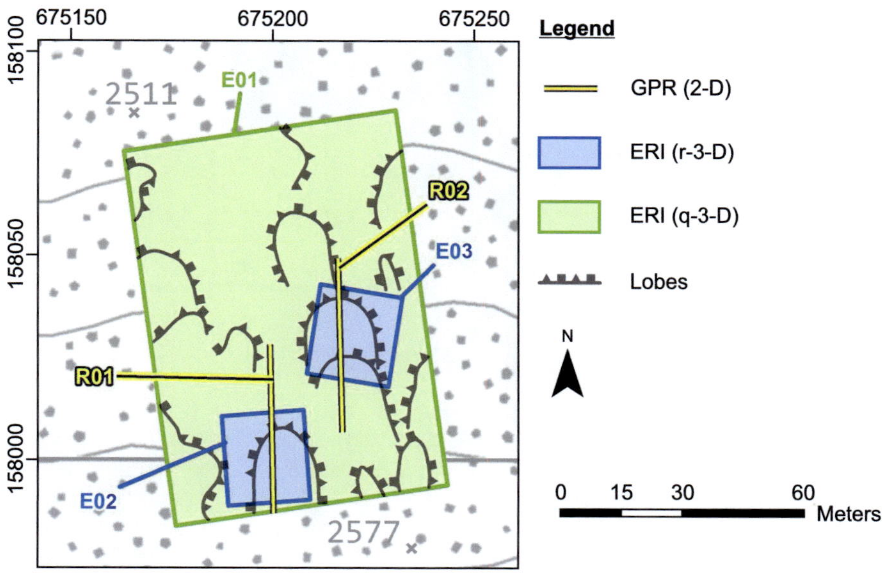

Basemap: Landeskarte 1:10'000 - Contour Line Spacing: 20 m
Coordinate System: CH1903 LV03 - Reproduced by Permission of swisstopo (BA19100)

Figure 3.10: Map of Survey Locations FUR.

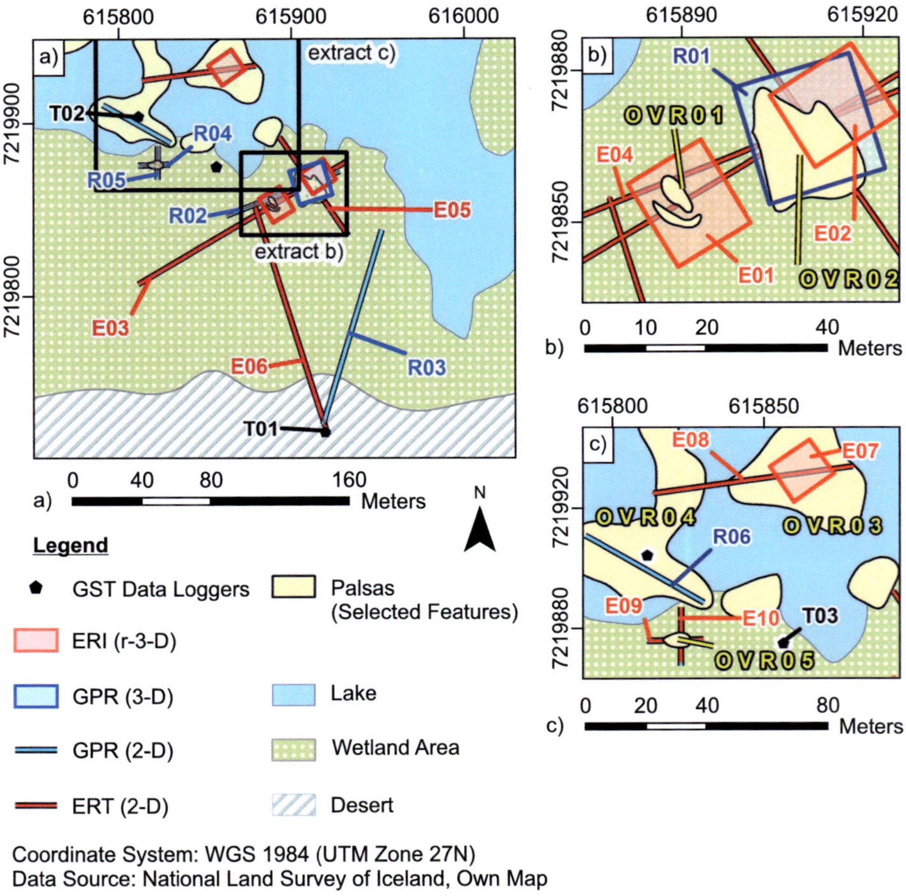

Figure 3.11: Map of Survey Locations OVR.

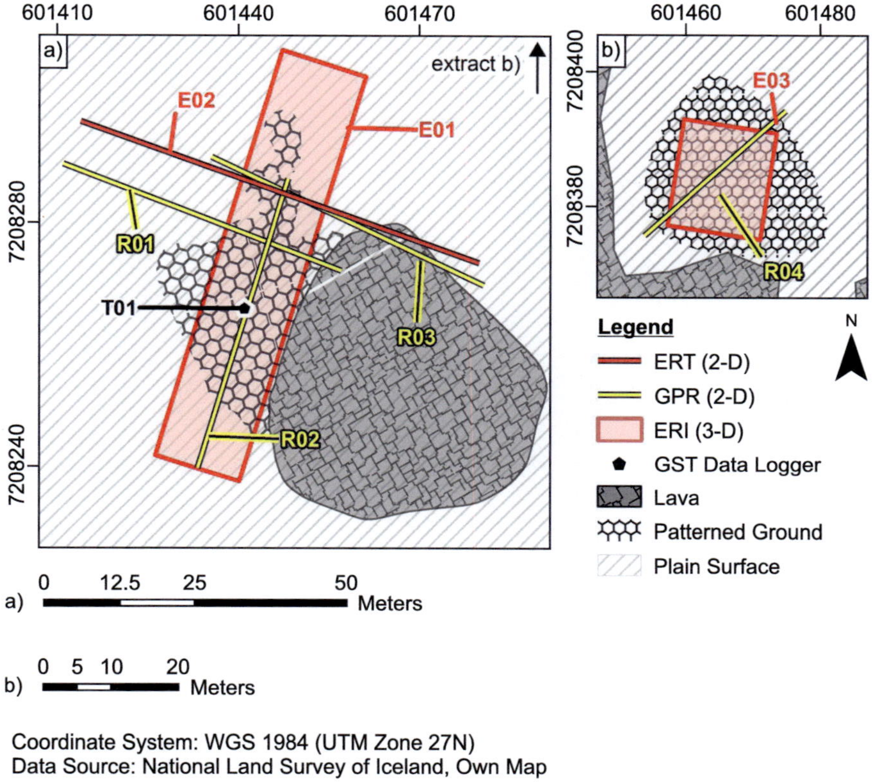

Figure 3.12: Map of Survey Locations HPA.

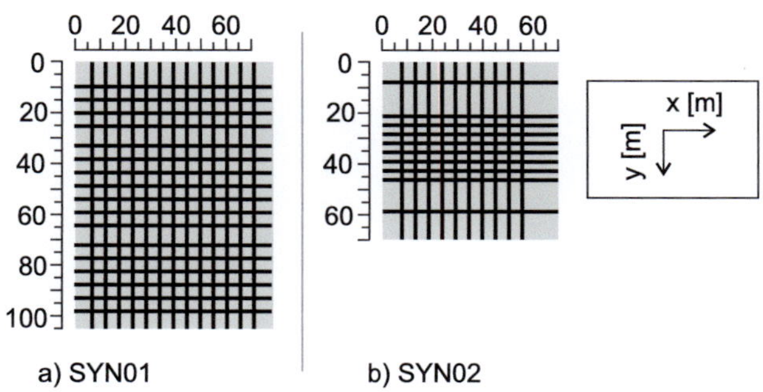

Figure 3.13: Sketches of Survey Lines SYN. a) SYN01, b) SYN02.

Chapter 4

Results and Interpretation NAR

Geophysical investigations at NAR comprise multi-dimensional approaches of ERI (4.1) and GPR (4.2). Temperature data was recorded at the ground surface and by a subsurface thermistor chain (4.3). The last part of this chapter comprises an interpretation of the results, focusing the internal structures of the rock glaciers NAR01 and NAR07 as well as the general permafrost distribution and ground ice characteristics of the study area (4.4).

4.1 Electrical Resistivity Imaging

Presented results of geoelectrical surveying are divided into three parts: Surveys targeting rock glacier NAR01 are presented in the first part of this section, while surveys targeting rock glacier NAR07 are presented in the second part. The third part of this section comprises results from six reconnaissance surveys conducted on the upper lobes of the rock glacier assembly, and from eight reconnaissance surveys conducted on the lower lobes. A previous comparison between results of 2-D ERI surveying, 2-D SRT surveying and borehole temperature measurements (Emmert and Kneisel, 2017) shows that a value of around 7 kΩm can be used to distinguish between frozen and unfrozen subsurface conditions at NAR01 rock glacier. This threshold value is used for interpretation purposes of the results from all rock glacier lobes, as surface conditions seem comparable. Although results of survey E01 have already been published (Emmert and Kneisel, 2017), they are rearranged here, as the additionally performed measurements considerably enhance the interpretation capabilities.

Rock Glacier NAR01

The most westerly located rock glacier NAR01 reaches the highest elevation of all investigated rock glaciers. It was investigated by one q-3-D ERI survey (E01), two one-year periods of ERTM surveying (E02, E03) and two supplementary 2-D ERI surveys (E04, E05). Both periods of ERTM surveying were processed independently to eliminate the influence of inter-annual variations, like temporal shifts in the snowmelt period. Model E02 comprises data acquired in 2015 and model E03 comprises data acquired in 2016. When qualitative changes of resistivity values between the different time steps (TS) are described, median values are used due to their lower sensitivity to single outliers. Additionally, the number of cells which are concerned is stated in brackets.

In the upward part of the q-3-D ERI model E01 (fig. 4.1), resistivity values between 200 kΩm and 400 kΩm indicate ice-supersaturated conditions in an area that corresponds to the talus slope behind the rock glacier (A). At this spot A, the extremely high resistivity values form a band of 40 m width in x-direction, which is visible throughout the entire

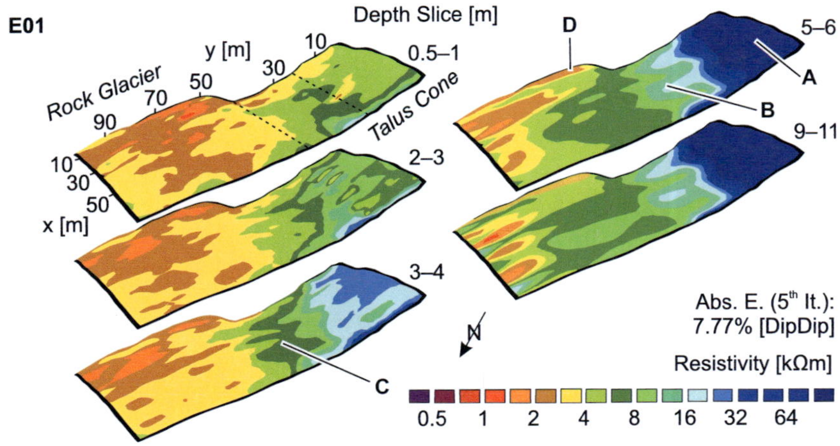

Figure 4.1: ERI Model NAR: E01. Selected Slices. Labels Refer to Structures Mentioned in the Text.

model depth of 15 m. In y-direction, resistivity values decrease to a range between 12 kΩm and 20 kΩm (B). Values of this range still indicate frozen conditions but indicate a lower ice content. The part of the model around spot B corresponds to the root zone of the rock glacier. It is covered by a layer with around 2 m thickness and resistivity values between 4 kΩm and 7 kΩm. The observed decrease in resistivity continues in y-direction, while the thickness of the upper layer simultaneously increases to around 4 m (C). Resistivity values increase in x-direction towards the western margin of the model, while relatively low resistivity values (<4 kΩm) are present throughout the entire model depth at the eastern margin (D).

The ERTM model E02 repeats the appearance of the prominent structures of the E01 model (fig. 4.2): resistivity values of up to 400 kΩm appear within the first 20 m of the profile (E), in a section which corresponds to the talus cone. The rest of the model is vertically divided into two parts: an upper layer, in which resistivity values below 7 kΩm indicate unfrozen conditions (F), and a lower layer, in which resistivity values above 7 kΩm indicate frozen conditions (G). However, the position of the boundary between these two layers F and G varies in space and time: Its minimum depth of around 1.3 m is reached at $x = 10$ m and remains constantly at this depth level throughout the complete investigation period. The maximum depth of the boundary is around 6 m and is reached at $x = 49$ m, but only in TS5.

Within the first 20 m of the profile, the observed resistivity changes differ between the model cells, dependent on their initial resistivity value. Cells with an initial resistivity value above 50 kΩm ($n = 6$), which is assumed to represent extraordinary ice rich conditions in this case, show a median increase of 53 % from TS1 to TS2, while cells with an initial value between 50 kΩm and 20 kΩm ($n = 15$), which represents a lower ice content, show a median decrease of 22 %. From TS2 to TS3, resistivity values decrease in all cells with an initial value between 20 kΩm and 50 kΩm, but the decrease is stronger in the cells with an initial value above 50 kΩm. While resistivity changes are neglectable between TS3 and TS4, resistivity values of all cells in this part of the model increase strongly towards TS5. The increase is 26 % for cells with an initial resistivity value above 50 kΩm and 47 % for cells with an initial resistivity value between 50 kΩm and 20 kΩm.

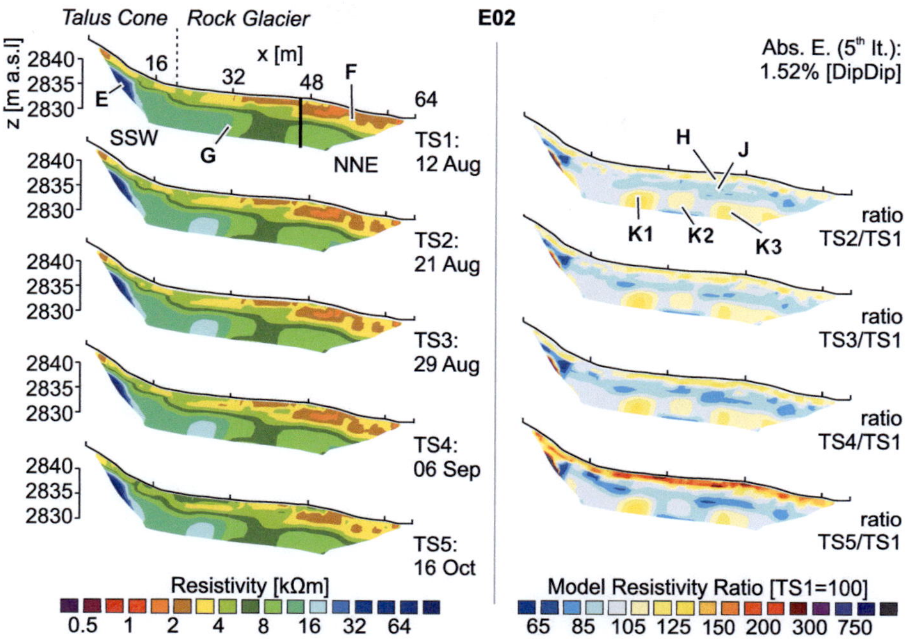

Figure 4.2: ERI Model NAR: E02. Labels Refer to Structures Mentioned in the Text. Vertical Line Marks Borehole Position.

The adjacent part of the model corresponds to the actual rock glacier. In this part, time-lapse results show a vertical subdivision at a depth of around 1.7 m. It divides the upper model layer F that, according to the assumed threshold value (7 kΩm), represents the active layer. Resistivity values in cells above this boundary (H) increase with a median of 12 % from TS1 to TS2 ($n = 143$). Values remain on this higher resistivity level and show only small fluctuations (± 5 %), until the resistivity values increase again by 28 % from TS4 to TS5. Model cells below the boundary (J) ($n = 211$), show a median decrease of 8 % from TS1 to TS2, followed by alternating resistivity changes between 6 % and -5 %. Model cells with resistivity values permanently above 7 kΩm ($n = 153$) are assumed to represent permafrost. Affected cells show a median resistivity increase of 8 % between TS1 and TS2 and a constant decrease in the following. The initial increase is particularly strong at three spots (K1–K3), while resistivity values of the other cells remain on a nearly constant level. From a total of 654 model cells, resistivity values of 47 cells shift across the threshold value. This transition affects (i) parts of the shallow subsurface, in which cells exceed the threshold value between the second last and the last time step ($n = 6$), but also (ii) model cells at the bottom of the active layer with initially low resistivity values that vary around the threshold value in alternating directions ($n = 41$).

The ERTM model E03 (fig. 4.3) shows the same structures as the E02 model and therefore the same labeling is used. Prominent differences between the models concern the lower layer (G), where resistivity values at TS1 (up to 21 kΩm) are higher in the E03 model compared to the E02 model (up to 19 kΩm). The depth of the boundary between the upper layer of low

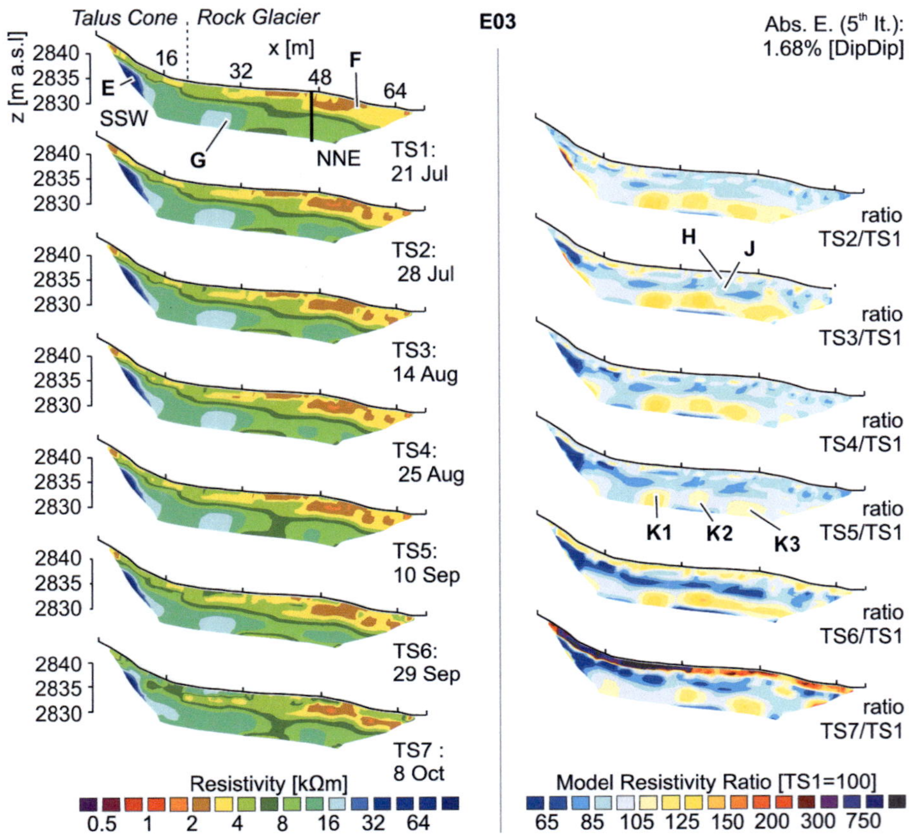

Figure 4.3: ERI Model NAR: E03. Labels Refer to Structures Mentioned in the Text, Vertical Line Marks Borehole Position.

resistivity (F) and the lower layer of high resistivity (G) varies in a similar way as observed in the E02 model. Additionally, the magnitude of the observed resistivity changes differs between the models. This concerns the part of the model that corresponds to the talus cone (E), in which the contrasting resistivity development between the cells with high and low initial resistivity values is less pronounced in the E03 model. After an initial median increase of 43 %, resistivity values of model cells with initial resistivity values above 50 kΩm ($n = 6$) decrease until TS6 and increase afterwards. Model cells with initial resistivity values between 20 kΩm and 50 kΩm ($n = 15$) show constantly decreasing resistivity values, which decline by median values between 4 % and 24 %.

In the part of the model that corresponds to the actual rock glacier, the upper (H) ($n = 143$) and the lower part of the active layer (J) ($n = 202$) show an initial decrease of 7 % ($n = 143$). Hence, the contrasting resistivity development between these two layers, which is observed at corresponding time steps of the E02 model, is not yet present in the E03 model. However, a contrasting development is visible between the following time steps, when values in the upper part (H) increase by a median of 8 %, while values in the lower part (J) decrease

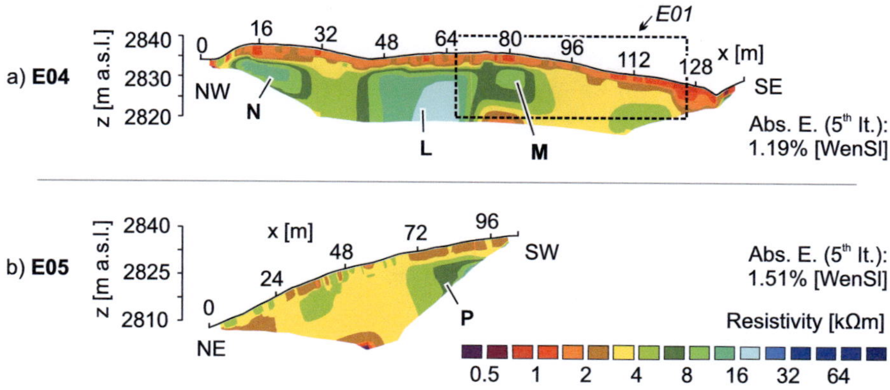

Figure 4.4: ERI Models NAR: a) E04, b) E05. Labels Refer to Structures Mentioned in the Text, Dashed Box Indicates Overlapping Section with E01.

constantly until TS6. Values in the upper part decrease slightly between TS3 and TS5 but increase again by 17 % between TS5 and TS6, and by 43 % between TS6 and TS7. In the lower part of the active layer, values increase by 5 % between TS6 and TS7. Similar to the observation in the E02 model, the E03 model shows increasing resistivity values between TS1 and TS2 in cells that represent permafrost conditions ($n = 162$). This increase is followed by slightly decreasing resistivity values throughout nearly the rest of the investigated time span; only between TS5 and TS6, the decrease is interrupted by stable conditions (±3 %). From 654 model cells, 125 cells shift across the threshold value in the E03 model. This number is much higher compared to the E02 model. It is attributed to cells of the shallow subsurface that exceed the threshold between TS06 and TS07 ($n = 87$). The number of model cells that shift across the threshold at the bottom of the active layer (F), is only slightly different between E02 and E03 ($n = 38$).

Two additional 2-D ERI surveys stretch the investigated area of NAR01 in transverse (E04) and longitudinal (E05) direction, and provide additional information on deeper layers. The resistivity distribution of the E04 model (fig. 4.4a) broadly repeats the structure of the overlapping part of the E01 model. Minor differences are attributed to seasonal or interannual alterations between the survey dates (see tab. A.4 and tab. A.1, appendix). Adjacent to the structure of high resistivity in the overlapping part (M), the E04 model shows an additional structure of high resistivity (L). Resistivity values of structure L exceed the values of structure M and increase with depth from 8 kΩm to 20 kΩm.

Resistivity values above 7 kΩm in the area between the two structures of relatively high resistivity indicate that the two structures L and M are part of one frozen compound structure. Beside this central area of high resistivity, a third structure of relatively high resistivity values around 14 kΩm appears at the northwestern end of the survey line (N). Model E05 (fig. 4.4b) overlaps with the E02/E03 models on a length of 29 m. It shows that the permafrost layer, as derived from the E02/E03 models, ends abruptly behind the overlapping section (P). The downwards following part of the model, which is characterized by a slightly steeper surface topography, is dominated by relatively low resistivity values (<5.5 kΩm) and thus indicates unfrozen conditions.

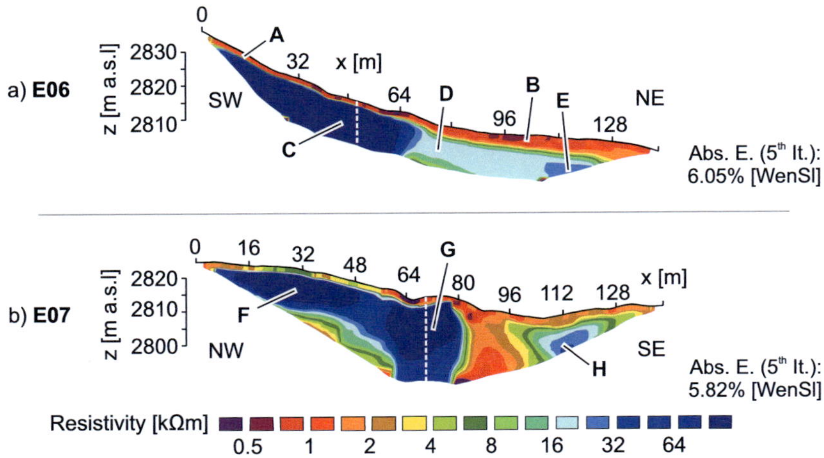

Figure 4.5: ERI Models NAR: a) E06, b) E07. Labels Refer to Structures Mentioned in the Text, Dashed Lines Indicate Intersection Between the Models.

Rock Glacier NAR07

Rock glacier NAR07 is much smaller than the other rock glaciers of the assembly and superimposes the root zone of the rock glaciers NAR03 and NAR05. The rock glacier was investigated by two 2-D ERI surveys (E06, E07) and two ERTM surveys (E08, E09).

The resistivity distribution of the longitudinal E06 model (fig. 4.5a) is qualitatively similar to the resistivity distribution of the longitudinal models of NAR01 rock glacier, but resistivity values are much higher in the E06 model. An upper layer with resistivity values below 7 kΩm indicates unfrozen conditions and shows an increasing thickness in a downslope direction, from 1.5 m (A) to 4.5 m (B). Extremely high resistivity values below this layer (up to 850 kΩm) indicate ice-supersaturated conditions at a central position (C). While resistivity values in the upper part of the profile are only slightly lower, resistivity values drop to 17 kΩm in a downslope direction (D). This drop is followed by a gradual increase in resistivity to values up to 23 kΩm (E). The range of resistivity values at the two spots D and E indicates frozen conditions but different ground ice characteristics, compared to spot C.

The transverse E07 model (fig. 4.5b) shows relatively high resistivity values at three spots (F, G, H). Extremely high resistivity values (>700 kΩm) appear at positions corresponding to the root zone of the adjacent rock glacier NAR02 (F) and to the upper lobe of NAR07 (G). They indicate ice-supersaturated conditions. Resistivity values of up to 35 kΩm at the third spot (H), which corresponds to a part of the talus slope adjacent to NAR07, also indicate frozen materials, but with lower ice content. Which a thickness of up to 5 m, the covering layer of relatively low resistivity is distinctly thicker above this spot H, compared to the rest of the model, where it reaches around 1.5 m. Resistivity values above 55 kΩm between the spots F and G indicate the existence of a continuous frozen layer with a minimum thickness of 12 m. Between the spots G and H, resistivity values decrease to 1.9 kΩm. This indicates unfrozen conditions throughout the entire model depth at this position, which corresponds to the transitional area between NAR07 and an adjacent talus slope.

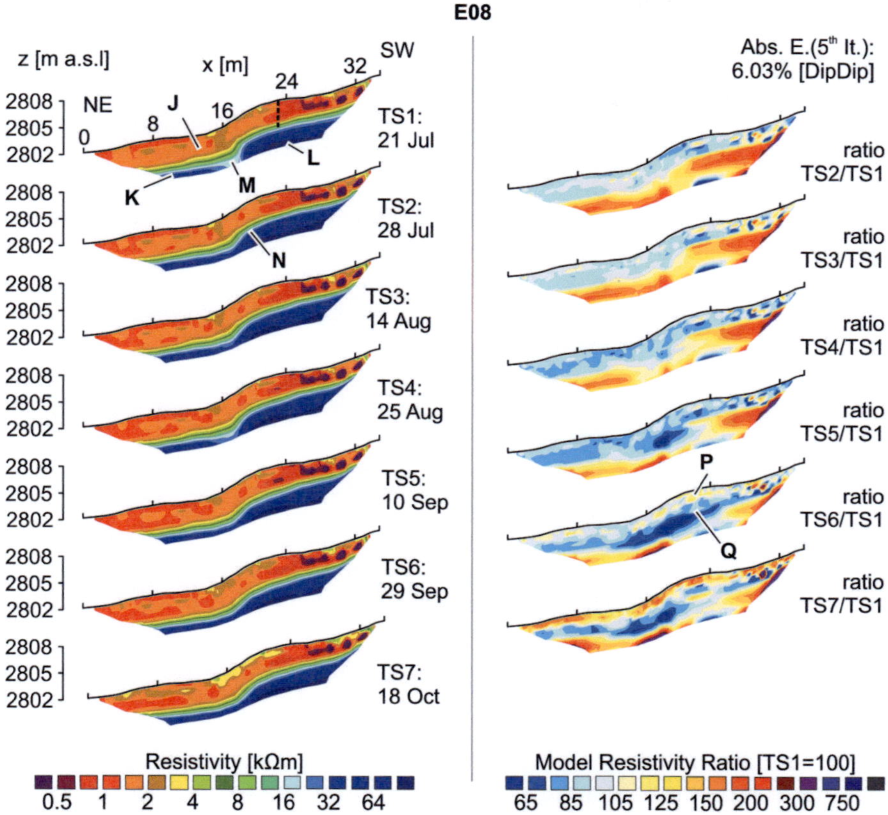

Figure 4.6: ERI Model NAR: E08. Labels Refer to Structures Mentioned in the Text, Dashed Line Indicates Intersection with E09.

The longitudinal ERTM model E08 shows two clearly separated layers (fig. 4.6). Resistivity values mainly below 4 kΩm indicate unfrozen conditions in an upper layer (J). This layer covers an area in which extremely high resistivity values indicate frozen and ice-rich conditions. While maximum resistivity values of up to 500 kΩm cluster at both ends of the survey line (K, L), resistivity values between these two spots are markedly lower (M). The thickness of the upper layer varies in space and time: its minimum thickness of around 1.7 m is reached at a $x = 31.5$ m in TS2 and TS5, and its maximum thickness of around 3.7 m is reached at $x = 15.5$ m in TS7. The position of the lower boundary of this upper layer J rises throughout the complete length of the profile by up to 0.8 m between TS1 and TS2. It subsequently descends with variable intensities throughout the subsequent four time steps. From TS6 to TS7, the position of the boundary rises only slightly at the most upward part of the profile but remains at a constant depth level throughout the rest of the profile. The observed differences in the intensity of the descend between TS2 and TS6 lead to a smoothing of the previously more angularly shape of the boundary, and to a downslope shift of the prominent step of its outline (N) below the front of the upper lobe.

When the threshold value of 7 kΩm is used for a segmentation of the resistivity distribution of the E08 model, a median decrease in resistivity of 6 % affects the unfrozen part of the subsurface between TS1 and TS2 ($n = 496$). After only slight alterations between TS2 and TS3, resistivity values decrease by a median of 7 % from TS3 to TS4. After TS4, the active layer is vertically subdivided at a depth of around 1 m. Resistivity values above this boundary (P) ($n = 258$) remain at a nearly constant level between TS4 and TS5, while values below this boundary (Q) ($n = 238$) continue decreasing with a median of 6 %. The decrease in this lower part of the active layer continues from TS5 to TS6, while values in the upper part of the active layer start to increase. From TS6 to TS7, resistivity values in both parts increase by a median of 34 % (P) and 6 % (Q), respectively. Model cells with an initial resistivity value above 7 kΩm ($n = 158$), increase by a median of 52 % from TS1 to TS2. After TS2, resistivity values decrease by median values between 3 % (TS2 to TS3) and 10 % (TS5 to TS6). This decrease in resistivity compensates the initial increase at 63 of the 158 model cells. Model cells that are affected by this compensation are mainly located in the part of the model that corresponds to the area around the front of the upper lobe. Towards TS7, resistivity values increase again by a median of 9 %.

The transverse E09 model (fig. 4.7) reflects only the upper part of the E08 model, while the lower part is not depicted due to a lower investigation depth. The resistivity distribution shows an undulating course of the boundary between an upper layer of low (R) and a lower layer of high (S) resistivity values. Resistivity values of the upper layer R agree well with values of the corresponding layer J of the E08 model. However, maximum resistivity values of the E09 model reach only up to around 60 kΩm. Values of this range cluster at two spots at the bottom of the model (T, U). In addition to the structures repeated from E08, the resistivity distribution of E09 shows a nearly continuous, surface-parallel band of relatively high resistivity values between 4 kΩm and 12 kΩm. It appears within the upper layer R at a depth of about 0.5 m (V).

Between TS1 and TS2 of E09, the boundary between the layers of low (R) and high (S) resistivity rises in a similar way as observed in the E08 model. The modeled resistivity distribution also repeats the period of stagnation between TS2 and TS3. From TS3 to TS6, the depth level of the boundary descends gradually by a summarized descend of around 0.2 m, before a slight rise is observed in some parts of the model between TS6 and TS7. Model cells that represent permanently frozen conditions ($n = 28$) show a particularly strong increase by a median of 95 % between TS1 and TS2. From TS2 to TS6, resistivity values decrease by median values between 2 % (TS2 to TS3) and 37 % (TS5 to TS6). This decrease overcompensates the initial increase at nearly all model cells. Between TS6 and TS7, all model cells with values permanently above 7 kΩm show a median increase of 25 %.

Model cells that represent the active layer show a rather fragmented pattern of resistivity changes. Cells with an initial value below 1.5 kΩm ($n = 318$), which appear in the uppermost 0.4 m of the model and again between depths of 0.8 m and 1.7 m (W), decrease by a median of 10 % between TS1 and TS2. In contrast to this decrease, model cells with an initial value between 1.5 kΩm and 7 kΩm (V) ($n = 308$), increase by a median of 9 %. This contrasting development weakens towards TS3, but enhances subsequently, when cells with a low initial resistivity value decrease by a median of only 3 % while cells with a relatively high initial resistivity value decrease by a median of 12 %. Between TS5 and TS7, both parts of the upper layer show a similar median decrease.

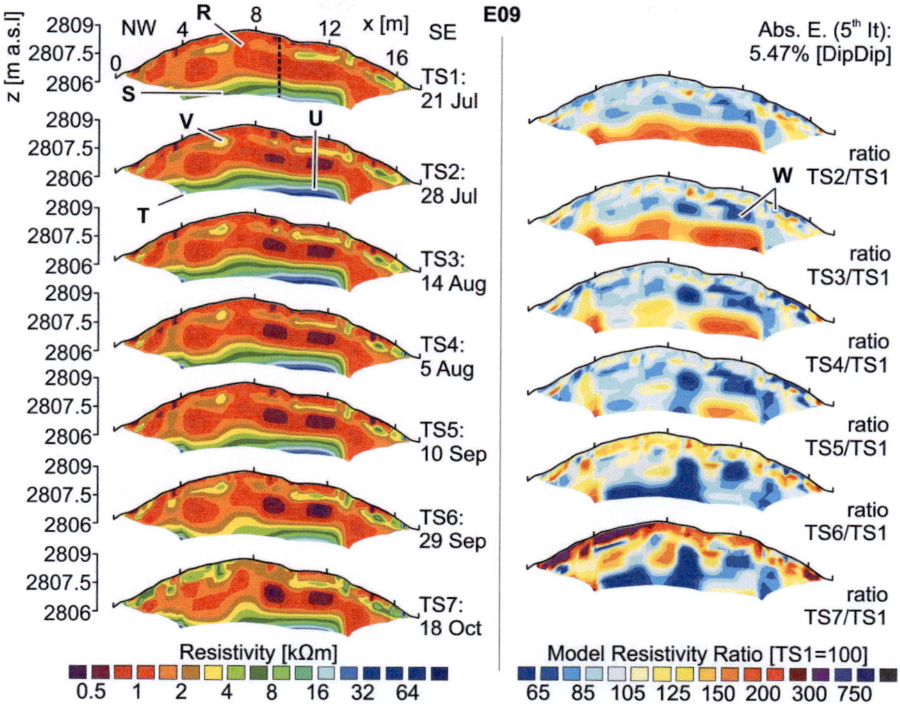

Figure 4.7: ERI Model NAR: E09. Labels Refer to Structures Mentioned in the Text, Dashed Line Indicates Intersection with E08.

Surveys on Upper and Lower Lobes

Six 2-D ERI surveys were performed at NAR01, NAR02 and NAR05 to investigate the subsurface conditions of the upper lobes and the transitional areas between the rock glaciers. The lower lobes of NAR02, NAR03, NAR04 and NAR05 were investigated by longitudinal, transverse and oblique 2-D ERI survey lines. At NAR06, one q-3-D ERI survey was performed. Core drilling, down to a depth of 4.5 m, was additionally performed at NAR06.

The resistivity distribution of the E10 model (fig. 4.8a) broadly reflects the resistivity distribution of the models E01 and E04 in the overlapping parts. Differences concern the magnitude of resistivity at structure B, which is 13 kΩm in the E04 model but 40 kΩm in the E10 model and the shape of structure B, which is rather blocky in the E10 model. In addition to the two prominent structures of the overlapping part (A, B), a third structure of relatively high resistivity appears in the E10 model: values of up to 55 kΩm indicate ice-rich conditions at the northwestern margin of NAR02, which is separated from NAR01 by a morphological trench of around 3 m depth (C). The topographic position of structure C is similar to the position of structure B, which is located at the northwestern margin of NAR01. Furthermore, the E10 model shows a large, triangular structure of extremely low resistivity values between 0.5 kΩm and 1.5 kΩm (D) at the bottom of the model. These values indicate unfrozen conditions and a high amount of liquid water.

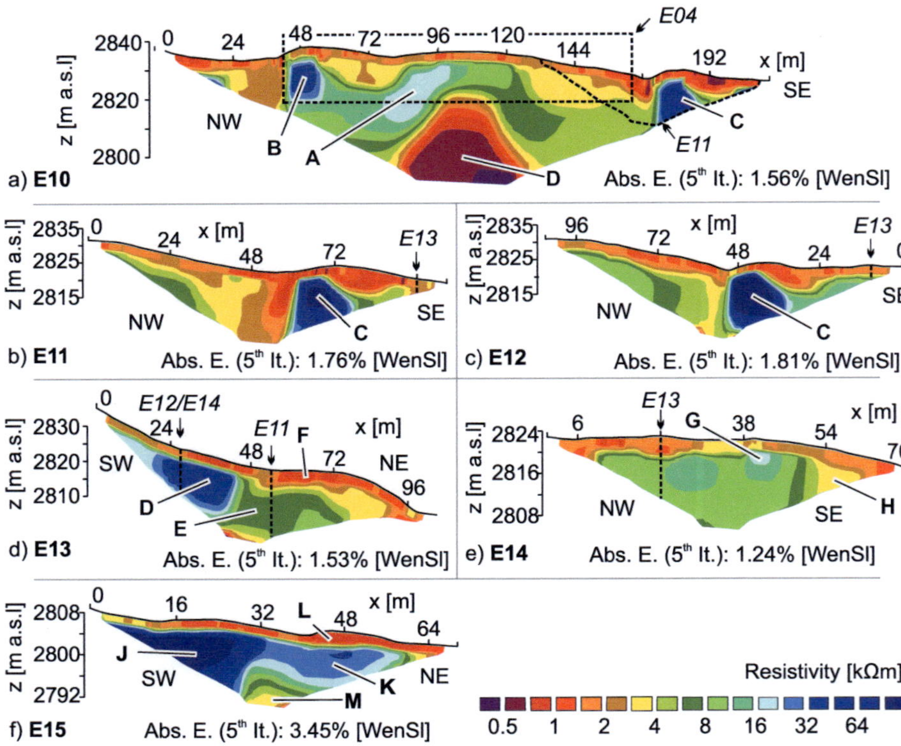

Figure 4.8: ERI Models NAR: E10–E15. Labels Refer to Structures Mentioned in the Text. Dashed Lines Indicate Intersections, Dashed Boxes overlapping Areas.

The E11 model (fig. 4.8b) overlaps the E10 model nearly completely, but surveying was performed more than two years earlier. Hence, a strong impact of interannual alterations is assumed. The two models show markedly different resistivity values for structure C in the overlapping part: while values of the E11 model reach up to 100 kΩm, values in the E10 model reach only up to 55 kΩm. A comparison between the two parallel models E11 and E12 (fig. 4.8c) shows that the spatial extent of structure C is constant in longitudinal direction of the rock glacier. However, maximum resistivity values of the structure decrease strongly, from nearly 170 kΩm in the E11 model to around 100 kΩm in the E12 model. This indicates a reduced ground ice content. Except for this difference at structure C, the E11 model shows lower resistivity values in nearly all other parts of the model.

The models E13 (fig. 4.8d) and E14 (fig. 4.8e) represent cross-sections of the root zone of NAR02. The longitudinal E13 model shows a resistivity distribution qualitatively similar to the resistivity distribution observed at the adjacent rock glacier NAR01 (E02, E03). Maximum resistivity values below the talus cone (D) are followed by markedly lower resistivity values in a downslope direction (E). The layer of relatively high resistivity values is covered by a layer of relatively low resistivity values (F). In a downslope direction, the thickness of this upper layer F increases. Maximum resistivity values of the E13 model reach 57 kΩm. This is distinctly lower than the maximum values of the E02/E03 models, which reach 400 kΩm.

The transverse E14 model overlaps the E12 model on a length of 31 m but surveying was performed nearly two years later (see tab. A.4, appendix). The E14 model shows a relatively homogenous resistivity distribution. Extensive parts of the model show resistivity values between 7 kΩm and 15 kΩm and indicate frozen subsurface conditions. A thickness of the active layer between 2 m and 4 m is assumed. Only at one spot (G), which corresponds to the southeastern margin of NAR02, resistivity values increase to 19 kΩm and thereby indicate a higher ice content. At the end of the survey line, where the topography starts to decline, resistivity values below 5 kΩm indicate an abrupt end of the frozen subsurface layer (H). At the intersection of the models E13 and E14, resistivity values differ strongly. While the longitudinal E13 model shows maximum resistivity values of 55 kΩm at depths below 5 m, maximum values of the transverse E14 model are only 12 kΩm in the same depth range. Further differences are indicated at the intersection of the models E11 and E13 and at the intersection of the models E12 and E13. However, the shallow investigation depths of the transverse models at the intersecting positions disables a detailed analysis.

The E15 model (fig. 4.8f) corresponds to a cross-section of the root zone of rock glacier NAR05 and shows a nearly continuous layer of relatively high resistivity values. At two positions, values of 130 kΩm (J) and 34 kΩm (K), respectively, indicate ice-rich conditions. Values in between the two spots J and K are around 26 kΩm. This still indicates frozen conditions but a lower ice content. The observed resistivity distribution qualitatively resembles the distribution of the parallel E06 model, but the range of resistivity values is smaller in the E15 model. Relatively low resistivity values in the shallow subsurface (L) and in deeper parts of the E15 model (M) indicate that the frozen layer is embedded between two unfrozen layers. For the frozen layer, a thickness of around 10 m is assumed at spot J.

Resistivity values of the E16 model (fig. 4.9a), which corresponds to a longitudinal cross-section of rock glacier NAR02, are between 0.3 kΩm and 12 kΩm. Only eight of 662 model cells show values above the threshold of 7 kΩm and thereby indicate frozen conditions. These cells concentrate at a small patch in the upslope part of the model at a depth of around 6 m (A). The other parts of the E16 model and the complete extent of the adjacent E17 model (fig. 4.9b) show resistivity values below 7 kΩm and are therefore assumed to represent an unfrozen subsurface. This observation indicates that the occurrence of frozen conditions at NAR02 is restricted to the two upper lobes. Extremely low resistivity values (<1 kΩm) at the bottom of the E17 model (B) indicate the presence of liquid in the subsurface of this part of the rock glacier, where a small spring exists near the front of the rock glacier.

The E18 model (fig. 4.9c) corresponds to a cross-section of the rock glaciers NAR03 and NAR04 and the transition between the two rock glaciers. Although the resistivity distribution shows a similar layering as the resistivity distribution of the E16 model, lower resistivity values indicate unfrozen conditions throughout nearly the complete extent of E18. Only at one spot, values exceed 7 kΩm (C). However, as this concerns only four cells at the margin of the model, and as the maximum resistivity is still only 7.6 kΩm, it is unclear if this structure indeed represents frozen conditions. Lowest resistivity values cluster in a continuous layer near the surface, which has a thickness of about 2 m (D) and stretches across the transition between the two rock glaciers (E). This transition is clearly reflected in the resistivity distribution by resistivity values of around 3 kΩm below the upper layer (E).

The resistivity distribution of the longitudinal E19 model, which corresponds to a cross-section of NAR04 (fig. 4.9d) is divided into two main parts. The upslope part (F) shows

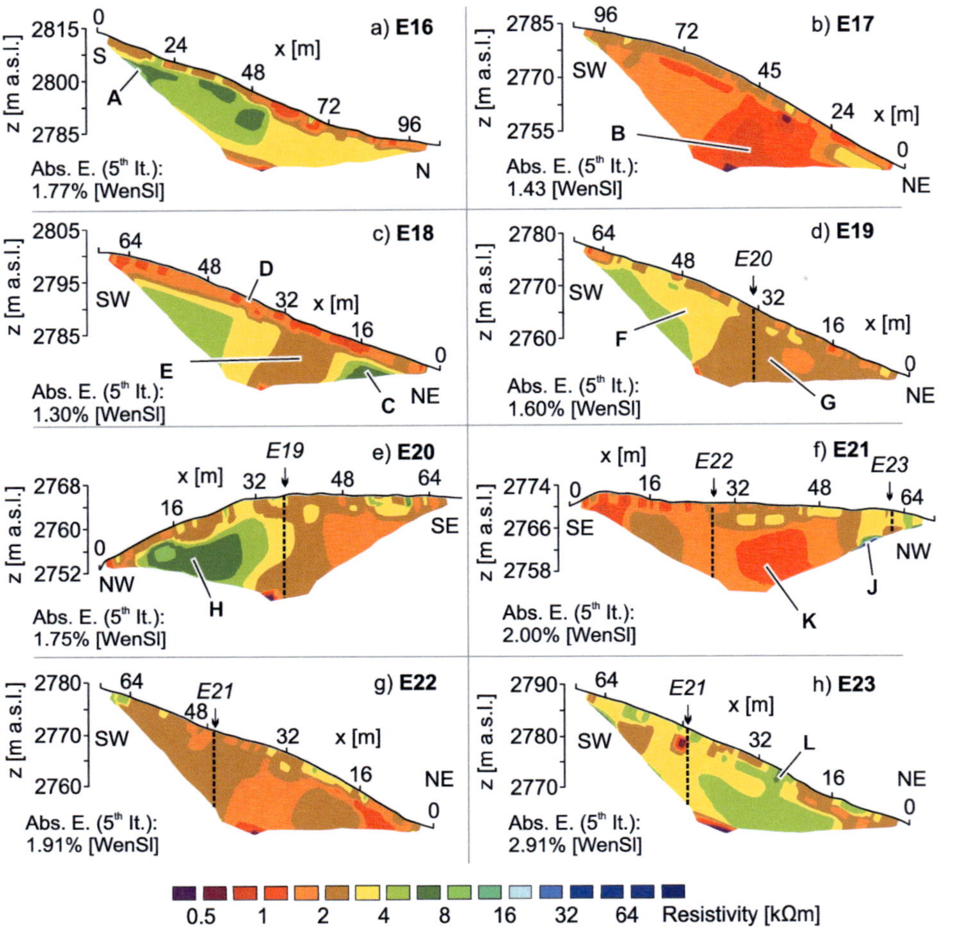

Figure 4.9: ERI Models NAR: E16–E23. Labels Refer to Structures Mentioned in the Text. Dashed Lines Indicate Intersections.

resistivity values between 3 kΩm and 5 kΩm, while resistivity values in the downslope part (G) are between 0.5 kΩm and 3 kΩm. The resistivity distribution indicates unfrozen conditions throughout the complete extent of the E19 model and an accumulation of liquid water in the downward part. In contrast to the models E16, E17 and E18, the E19 model lacks a continuous layer of relatively low resistivity values at shallow depths. The E20 model (fig. 4.9e) is perpendicular to the E19 model. At the intersection of the two models, resistivity values vary only slightly between 2 kΩm and 4 kΩm in both models and lack a distinct subsurface layering. Maximum resistivity values of the E20 model reach up to 8.6 kΩm and concentrate in a lens-shaped structure in the northwestern part of the model (H). This structure is the only part of the E20 model where resistivity values indicate frozen conditions.

The E21 model corresponds to a transverse cross-section of NAR05 (fig. 4.9f) and shows a relatively homogenous resistivity distribution. Only at the northwestern margin of the

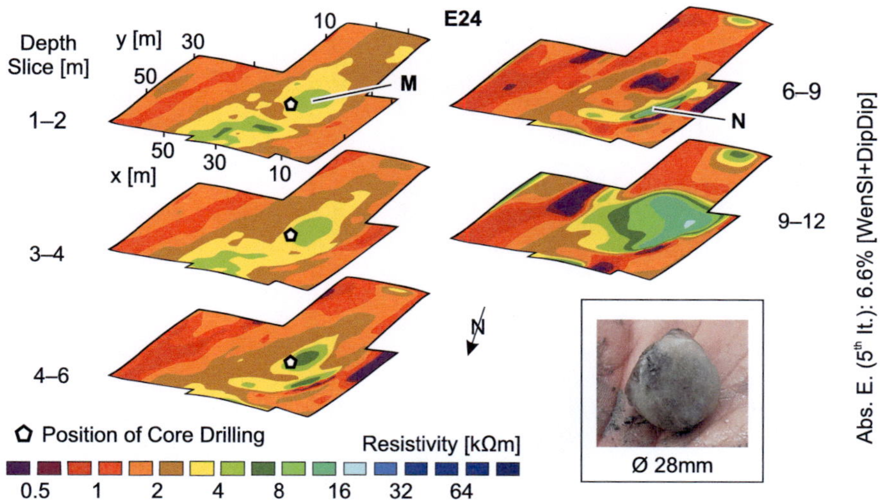

Figure 4.10: ERI Model NAR: E24. Labels Refer to Structures Mentioned in the Text.

model, anomaly high resistivity values (>100 kΩm) appear (J). This marginal structure is presumably caused by a bad ground coupling of the electrodes or by air-filled voids in the subsurface. Relatively low resistivity values in the deeper parts of the model (K) indicate the presence of liquid water. The E22 model corresponds to a longitudinal cross-section of NAR05 (fig. 4.9g) and agrees with the perpendicular E21 model on the observation of a relatively homogenous resistivity distribution. Values are below 5.5 kΩm throughout the complete extent of the model and thereby indicate unfrozen conditions. At the intersection of the models E21 and E22, values of the E22 model vary only slightly around 2 kΩm. This is similar to the values of the corresponding cells of the E21 model.

The resistivity distribution of the E23 model (fig. 4.9h) differs from the distributions of the other models of NAR05. It shows higher resistivity values, which reach up to 6.6 kΩm (L). Differences between the models E21 and E23 appear at the intersection of the two models: while the E21 model shows resistivity values of around 2 kΩm, corresponding cells of the E23 model reach values of up to 4 kΩm. In contrast to the resistivity distribution of the E18 model, which also spans across a transition between two adjacent rock glaciers, the transition between NAR03 and NAR04 is not reflected in the resistivity distribution of the E23 model.

Located at an elevation of around 2730 m a.s.l., NAR06 is the lowermost of the investigated rock glaciers. However, massive ground ice was detected at a depth of 4.4 m within the survey area of the E24 by core drilling on September 11, 2016 (fig. 4.10). The modeled resistivity distribution of the E24 model shows that relatively high resistivity values cluster in the central part of the model, where they form an irregularly shaped, longitudinal band with a width of about 34 m (M). This band is surrounded by values between 0.5 kΩm and 2 kΩm. Between depths of 1 m and 4 m, resistivity values change only slightly and reach up to 6 kΩm. This indicates that the entire survey area is ice-free at this depth level. Between depths of 4 m and 6 m, the horizontal extent of both structures decreases, but resistivity values increase to 8 kΩm and thereby exceed the threshold value of 7 kΩm. At depths between 6 m and 9 m,

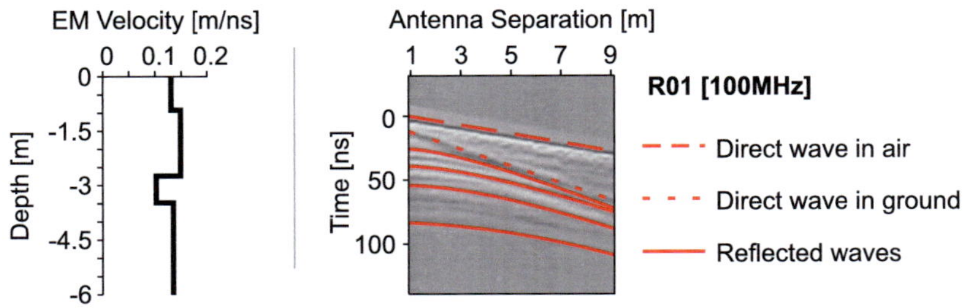

Figure 4.11: Radargram NAR: R01 (Right Panel) and Derived Velocity Model (Left Panel). Selected Reflectors are Accentuated in Red.

relatively high resistivity values disappear nearly completely and only a small structure of resistivity values up to 15 kΩm is visible (N). Below a depth of 9 m, resistivity values increase to 27 kΩm and relatively high resistivity values cover an extensive part of the model slice.

4.2 Ground-Penetrating Radar

In total, one CMP survey and two 2-D GPR surveys were performed at NAR01 and NAR07. As the previously presented results of the ERI models indicate distinct differences in subsurface conditions between the two rock glaciers NAR01 and NAR07, the vertical EM velocity layering of the CMP model was used for time-depth conversion of data from NAR01 only (R01). A constant EM velocity of 0.13 m ns^{-1}, based on values presented by Berthling and Melvold (2008), was applied for time-depth conversion of data from NAR07 (R02).

The R01 CMP model shows a structure of four layers (fig. 4.11). The uppermost layer has a thickness of 0.9 m and an EM velocity of 0.13 m ns^{-1}. It covers a layer with 1.8 m thickness and a velocity of 0.15 m ns^{-1}. Between depths of 2.7 m and 3.5 m, the EM velocity is 0.1 m ns^{-1}, and below this depth, the EM velocity increases to 0.13 m ns^{-1}. This layering indicates a relatively high amount of liquid water in the third layer, while the EM velocity values of the other layers are in a range that agrees with the assumed frozen conditions.

The R02 radargram (fig. 4.12a) shows steeply declining reflectors in the first part (A). In the following part, which corresponds to the transition between the talus cone and the rock glacier, it shows one distinct, slanting reflector directly below the surface (B). Below this reflector, a few undulating reflectors appear near the bottom of the radargram (C). One prominent reflector appears at a depth of around 3 m (D) in a section that corresponds to the upper lobe of NAR01. This slightly concave reflector is about 30 m long, and above and below this reflector, the radargram shows chaotic reflection patterns that predominately comprise short, curved reflectors. When the prominent concave reflector D starts to strike towards the surface, it is followed by a similarly shaped reflector (E). This reflector E itself is again followed by a third reflector (F), which is traceable throughout the rest of the radargram.

The R03 radargram (fig. 4.12b) is dominated by curved reflectors (G). In a part of the radargram that corresponds to the front of the upper lobe, undulating reflectors appear between 3 m and 5 m depth (H). A nearly continuous reflector is visible at a depth of around 3 m in a part of the radargram that corresponds to the lower lobe (J).

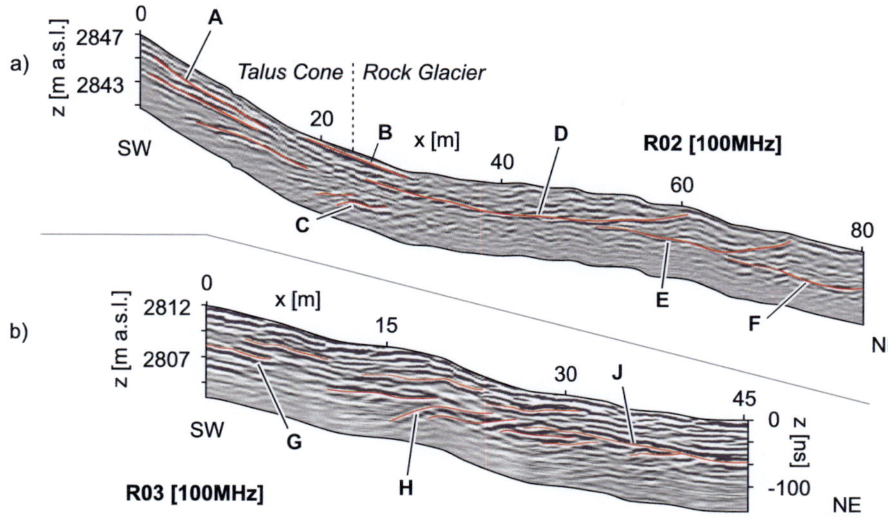

Figure 4.12: Radargrams NAR: a) R02, b) R03. Labels Refer to Structures Mentioned in the Text. Selected Reflectors are Accentuated in Red.

4.3 Temperature Measurements

Temperature data was continuously recorded by 15 temperature sensors, installed in a borehole of 10 m depth and at ten positions on the surface.

Subsurface Temperatures

Throughout the complete investigated time span, subsurface temperature values at NAR01 were at or below 0 °C at all sensors between 3 m and 5 m depth. However, the difference between the highest temperature and 0 °C is lower than the accuracy range of the sensors (±0.25 °C), and therefore the occurrence of permafrost cannot be validated. The same applies to the depth level between 5 m and 9 m, where temperature values between −0.01 °C and −0.25 °C were recorded. Only at depths below 9 m, temperature values were below −0.26 °C throughout the recorded time span and thereby verify permafrost conditions at NAR01.

The borehole temperature record (fig. 4.13) shows that downward freezing from the surface started on October 27, 2014 in the winter of 2014/2015 (2015/2016: Oct 20, 2016/2017: Oct. 08), seven days after the frost table started to rise. This order was reversed in the following winter of 2015/2016, when downward freezing started 14 days earlier than upward freezing. In the winter of 2016/2017, downward freezing reached the frost table before any rise was detected. In the winter of 2014/2015, the subsurface was completely frozen 23 days after the onset of the freezing period (2015/2016: 21 d, 2016/2017: 23 d), and remained in a frozen state for 199 days (2015/2016: 237 d, 2016/2017: 215 d). Thawing of the subsurface column took 49 days in the following spring of 2015 (2016: 40 d, 2017: 52 d) and reached a maximum thickness of 5 m on July 25, 2015 (2016: Aug 13, 2017: Jul 25).

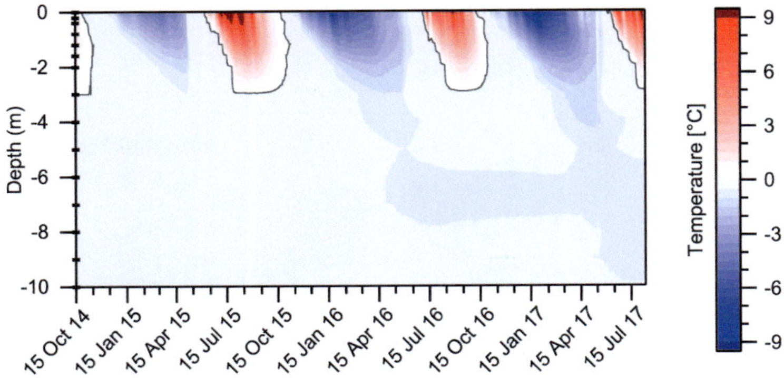

Figure 4.13: Recorded Borehole Temperature Data (NAR). Small Dots on y-Axis Indicate Sensor Positions.

The recorded subsurface temperature values show an increasing strength in the penetration of cold winter temperatures throughout the investigated time span. The wintry cooling of 2014/2015 caused a decrease in daily mean temperatures of 0.49 K at the sensor at 3 m depth, while temperature values below this level remained at a constant level. In the following winter of 2015/2016, the decrease in temperature reached down to a depth of 7 m, where temperature values decreased by 0.3 K. In the winter of 2016/2017, the seasonal decrease in temperatures reached down to the sensor at 9 m depth, where temperatures have been permanently below −0.5 °C since August 30, 2016.

Ground Surface Temperatures

The temperature records of the ten GST sensors, distributed over the rock glacier assembly, show MAGST values between −2.02 °C (T08: 2015/2016) and 1.81 °C (T10: 2014/2015) (tab. 4.1). Highest MAGST values appear at positions where the sensors are placed at concave surface positions like shallow depressions or trenches, followed by sensors at positions with a convex surface topography. Lowest MAGST values appear at a position in front of a lobe of NAR01 and at a position between coarse clasts in front of NAR01. This order correlates with the length of snow cover duration, as derived from the number of days with diurnal temperature variations below 0.4 K. At all sensor positions, MAGST values decrease between the two investigated periods. This is attributed to a period of extraordinary low GST values between August 2015 and January 2017. In turn, temperature values between September 2014 and July 2015 and after January 2017 were above the average.

The plots of the temperature data recorded from the data loggers at convex surface positions (T01, T02, T03) show highly fluctuating values and only short episodes of zero-curtain conditions (fig. 4.14). In the first year of the investigated period, they became snow-free at the beginning of June. This event shifted to the end of June/beginning of July in the following year of 2016. In 2017, they became snow free already at the end of May. While the GST sensors T02 and T03 changed their order of becoming snow-free first, the sensor T01 was the last one to become snow-free in all three years of the investigation. The three data loggers at convex surface positions recorded very low minimum daily mean temperature values of −11.21 °C (T01), −19.18 °C (T02) and −15.76 °C (T03), respectively.

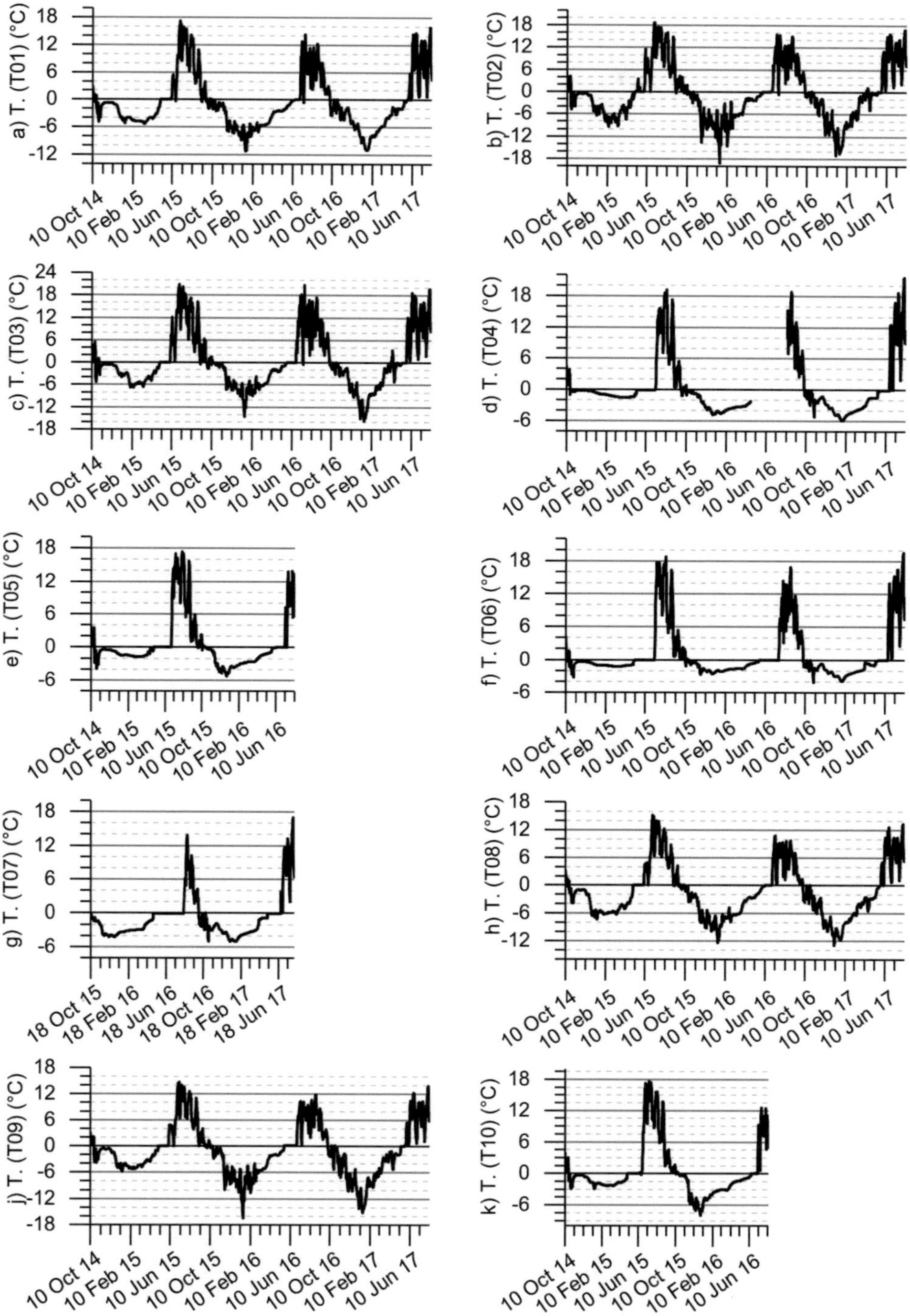

Figure 4.14: Recorded GST Data NAR.

Table 4.1: Summary of Recorded GST Data (NAR).

Label	MAGST [°C] 2014/2015[a]	MAGST [°C] 2015/2016[a]	SC [d] 2014/2015[a]	SC [d] 2015/2016[a]	SC [d] 2016/2017[b]	Topographic Position
T01	0.51	−1.4	224	180	171	Lobe Convex
T02	0.81	−1.19	121	100	65	Lobe Convex
T03	1.42	−0.25	220	139	86	Lobe Convex
T04	1.46	*No Data*	277	*No Data*	229	Lobe Concave
T05	1.77	*No Data*	260	*No Data*	*No Data*	Lobe Concave
T06	1.62	0.69	264	269	227	Trench Between Lobes
T07	*No Data*	-0.8	*No Data*	295	240	Lobe Concave
T08	-0.23	−2.02	224	185	134	Lobe Front
T09	0.43	−1.44	217	130	56	Coarse Clasts
T10	1.81	*No Data*	247	*No Data*	*No Data*	Coarse Clasts

MAGST = Mean Annual Ground Surface Temperature // SC = Snow Cover
[a]Investigated Period: 01 Nov–31 Oct // [b]Investigated Period: 01 Nov–09 Aug

The temperature plots of sensors placed at concave surface positions and in transitional areas between the rock glacier lobes show less fluctuations and more durable zero-curtain conditions. (fig. 4.14). The date on which the last of these sensors became snow-free was less variable during the investigated period and was always at the end of June/beginning of July, except for the sensor T07, which was covered with snow until mid-August in both years of recording. The occurrence of a persistent and insulating snow cover is assumed to limit the penetration of cold winter temperatures into the ground at these positions and therefore, the recorded minimum daily mean temperature values are relatively high. They are −5.90 °C (T04), −5.28 °C (T05), −4.16 °C (T06) and −5.03 °C (T07), respectively.

The plots of daily mean temperature values recorded at one of the lobe fronts of NAR01 (T08) and at one position between coarse clasts (T09) resemble the plots of the sensors placed at convex surface positions. Relatively low minimum daily mean temperature values of −16.39 °C (T08) and −12.97 °C (T09), respectively, were recorded. The temperature curve from the other position between coarse clasts in front of NAR01 (T10) shows similarities with the curves of the sensors at concave surface positions. However, recorded zero-curtain conditions are shorter and the MAGST is higher compared to these positions, although the recorded minimum daily mean temperature of −7.98 °C is relatively low.

4.4 Interpretation

In this section, the results of the different methods are jointly interpreted. It focuses on the internal structure of NAR01 and NAR07, but also addresses general characteristics.

Rock Glacier NAR01

The results of E04 and E10 affirm the assumed occurrence of ground ice in the root zone of NAR01, which is based on an interpretation of the E01 model by Emmert and Kneisel (2017). Differences in the resistivity distribution between the 3-D ERI model and the 2-D

ERI models are attributed to alterations caused by natural processes that occurred during the long time span between the survey dates (see tab. A.3, appendix). However, this argument cannot explain the differences between the E04 model and the E10 model, as surveying was performed on two subsequent days with stable weather conditions. Thus, differences between the two 2-D ERI models must be attributed to differences in the methodological approach. They may be caused by the spatial resolution of the two ERI models, which differs according to the different electrode spacings that were used (see tab. A.3, appendix). As a consequence, the generally lower data coverage of E10 is particularly reduced at marginal positions of the model. Nevertheless, results show a qualitatively similar characteristic resistivity pattern in lateral direction throughout the root zones of the rock glaciers NAR01 and NAR02.

Resistivity values of both models E04 and E10 agree on the occurrence of congelation ice from refreezing meltwater (Haeberli and Vonder Mühll, 1996). This meltwater presumably originates from snow or ice patches, which are buried under a shallow debris cover below the talus cones. It percolates slowly through the fine-grained debris and refreezes at undercooled parts of the rock glacier. The observation of longitudinal bodies of ground ice (e.g., E10–E12: C) indicates that the influence of a lateral meltwater flow is neglectable for the previously described way of ground ice formation, and that only meltwater that travels longitudinally through the rock glacier is affected by the described process of refreezing. The spatial coincidence between the abrupt end of the frozen layer, as observed in the longitudinal E05 model (E05: P), and an increasing slope angle of the surface topography indicates that the formation of lobes at NAR01 is connected to differences in ice content. The occurrence of buried patches of snow or ice in the subsurface of the talus cones is affirmed by the appearance of the strong, declining reflectors in the R02 radargram (R02: A). They are assumed to represent alternating layers of debris and massive ice (Moorman et al., 2003) and indicate that not one single, but multiple burial events must have occurred. The undulating reflectors in the lower part of the R02 radargram (R02: C) indicate a deformation by compressional flow and can be interpreted as a sign for active rock glacier movement (Degenhardt and Giardino, 2003; Monnier et al., 2008; Monnier et al., 2011).

The R02 radargram indicates the existence of a continuous frost table by the appearance of a prominent compound reflector (R02: D-E-F), which is traceable nearly through the complete radargram. The chaotic reflection pattern below indicates frozen conditions but also the absence of massive ice, as this part of the radargram lacks any blank spaces (Moorman et al., 2003). The upwardly striking parts of the compound reflector show the occurrence of thrusting processes (Monnier et al., 2013; Monnier and Kinnard, 2015). A boundary between an upper and a lower part of the active layer, as derived from the time-lapse models E02/E03, is not reflected in the R02 radargram, but the multi-layer stratification of the R01 CMP model affirms the existence of this feature. The therein observed layering aligns with the vertical resistivity gradient at a corresponding position of the time-lapse models, although the depths of the boundaries between the layers do not exactly correspond (not shown).

The observed intraannual variations in the resistivity distribution of E02/E03 indicate that melting starts at less ice-rich parts of the subsurface below the talus cone and that the resulting meltwater subsequently refreezes at parts with higher ice content. Only after TS2 (E02/E03), melting affects the complete subsurface of the talus cone. The initial increase in resistivity in the upper part of the active layer at NAR01 (E02/E03: H) is attributed to dryer conditions in this shallow part of the subsurface, due to a reduced supply with meltwater from surface snow. After TS2 (E02/E03), stable subsurface conditions establish, which last until a resistivity increase in the upper part of the active layer indicates the onset of wintry downward freezing at the last time steps of both periods. However, the observation of this onset disagrees with the borehole temperature records from the days of geoelectrical surveying in both years, as they show slightly positive temperature values at the affected depth level. This discrepancy is likely caused by small-scale differences in the ground thermal regimes between the borehole position and the location of the ERTM survey line, or by the weak accuracy of the temperature sensors. Recorded GST values from the sensor which was nearest to the survey line (T01) are slightly negative around the days of geophysical data acquisition.

The lower part of the active layer (E02/E03: J) is characterized by meltwater flow on the frost table during summer. This is indicated by the observation of relatively low resistivity values after the end of the snowmelt period, when meltwater accumulates on the frost table. A rise of the frost table between TS1 and TS2 is observed in both time-lapse models E02 and E03 and also reflected by a recorded decrease in borehole temperature values between the corresponding days. It indicates that percolating meltwater refreezes on the still relatively cold frost table. After this short period of refreezing ends, constantly decreasing resistivity values indicate a marginal melting at the top of the frost table. The therefore assumed impermeability of the frost table and ice-rich conditions align with the observed strong reflection amplitude of the compound reflector in the R02 radargram (R02: D-E-F). Small resistivity changes within the permafrost layer show that the influence of meltwater on this layer is rather low and restricted to small patchy structures with varying ratios between liquid water and ice (E02/E03: K1–K3). The last-named observation must be interpreted carefully, as resistivity contrasts in this part of the model may be enhanced unreasonably by the applied L1-norm inversion scheme or by a prolonged inversion process (Hauck and Vonder Mühll, 2003). However, the application of an L1-norm inversion scheme was necessary to avoid problems in detecting the lower limit of the frozen layer (Lewkowicz et al., 2011).

Small interannual differences between the time-lapse models E02 and the E03 are attributed to differences in weather conditions on the respective survey dates. The stronger resistivity increase in the upper part of the active layer from TS5 to TS6 in the E03 model is e.g., attributed to an extraordinary low precipitation in September 2016. In this month, a summarized precipitation of only 36 mm was recorded at the weather station in Samedan, which is half of the long-term average (MeteoSchweiz, 2018). The higher number of model cells that represent permanently frozen conditions at greater depths in the E03 model is attributed to an enhanced penetration of cold winter temperatures, which is also reflected by the borehole temperature records. The observed increase in resistivity in the upper part of the active layer between TS1 (E03) and TS2 (E03) is attributed to the earlier start of the investigation period (E02: 12 Aug; E03: 21 Jul) and different snowpack condition. On the first survey date of E02, the snowmelt period had nearly been completed at NAR, but the same position, in contrast, was still covered by about 1.8 m of snow on the first survey date of E03. This enormous snowpack had presumably bound a considerably amount of water, which was released only after TS2 (E03).

The strong resistivity decrease in the deeper parts of the E10 model (E10: D), may represent the lower boundary of the frozen layer, as its depth corresponds to the assumed permafrost depth that Emmert and Kneisel (2017) deduced from results of 2-D SRT surveying for this position. The extremely low resistivity values indicate a transition to unfrozen bedrock below the frozen layer (tab. 3.1). However, regarding the very low model resolution index values in this part of the model (see fig. B.1, appendix), it must be noted that the appearance of this structure is not necessarily backed by the data.

Rock Glacier NAR07

Qualitatively similar resistivity distribution patterns at NAR07 and NAR01 indicate an analogous way of rock glacier formation. Hence, the incorporation of surface snow or ice into the subsurface is also assumed to be crucial for the formation of ground ice at NAR07. Extremely high resistivity values (>850 kΩm) in the E06 model indicate ice-supersaturated conditions, but the lack of any characteristic reflection patterns in the E03 radargram, like e.g., blank spaces (Moorman et al., 2003), shows the absence of massive ice bodies in the subsurface.

The appearance of curved reflectors in the R03 radargram (R03: G) indicates that the upper part of the rock glacier consists of talus deposits and is affected by flow processes (Berthling et al., 2000). As flow processes are often associated with a state of water saturation (Benedict, 1976), the occurrence of an impermeable frost table is assumed, although the appearance of a corresponding, continuously traceable reflector is limited to the part of the radargram that corresponds to the lower rock glacier lobe (R03: J). Below this continuous reflector, the appearance of undulating reflectors in the R03 radargram (R03: H) is interpreted as a result of deformation by compressional flow (Monnier et al., 2011). The affected part of the radargram coincides with parts of the E06 and E08 model that show a sharp longitudinal drop in resistivity, which indicates a strong decrease in ice content.

Results of the E08 model show a strong increase in liquid water in the upper subsurface between TS1 and TS2. This reflects the substantial decrease in surface snow, which was observed in the field between the two dates of surveying. After the end of the snowmelt period around TS3 (E08), only slight variations appear in the active layer. The resistivity increase from TS5 to TS6 in the upper part of the active layer is again attributed to the extraordinary dry weather conditions in September 2016, as described previously (MeteoSchweiz, 2018). It is followed by an even stronger increase in resistivity in the upper subsurface between TS6 and TS7, which coincides with slightly negative GST values (T07) and is therefore interpreted as the onset of downward freezing. The assumed shift of the step-shaped part of the frost table in downward direction (E08: N) indicates a fast and intense movement of the upper lobe and aligns with the observed signs of compressional flow in the R03 radargram.

Relatively low resistivity values at a position that corresponds to the area between the upper and the lower rock glacier lobe (E08: M) indicate that liquid water accumulates at this position. This is supported by an assumed deepening of the frost table topography at this position, which develops throughout the investigated time span. This assumption aligns with the observation that the initial increase in resistivity is compensated relatively early in this part of the model, compared to other parts of the model, and that the following decrease in resistivity is particularly strong. A similar scenario can explain the resistivity changes in the E09 model: At a position with a slightly concave frost table (E09: T) resistivity values decrease constantly, but at a position with an inclining frost table (E09: U), resistivity values fluctuate after the end of the snowmelt period. The band-shaped structure of relatively high resistivity values in the shallow subsurface of the E09 model (E09: V) is interpreted as an interbedded layer of rather coarse-grained materials. This assumption is based on the stable

Table 4.2: Conversion Table: Potential Direct Solar Radiation (Jul–Oct) to BTS. Modified from Hoelzle (1992).

Pot. Dir. Sol. Radiation $[\mathrm{MJ\,m^{-2}\,d^{-1}}]$	Bottom Temperature of Snow Cover (BTS)
16.8	−8 °C to −7 °C
17.5	−7 °C to −6 °C
18.2	−6 °C to −5 °C
19.0	−5 °C to −4 °C
19.7	−4 °C to −3 °C
20.4	−3 °C to −2 °C
21.2	−2 °C to −1 °C
22.0	−1 °C to 0 °C

depth level of this structure throughout the investigated time span and the relatively strong absolute changes in resistivity, as this is characteristic for coarse-grained materials, due to their lower water retention capacity (Schneider et al., 2013).

Permafrost Distribution and Ground Ice Characteristics

Although the geophysical models show a widespread permafrost distribution at NAR that agrees well with maps of the potential permafrost distribution (BAFU, 2005; Boeckli et al., 2012), the presented GST plots show ambiguous results. Compared to the plots of the classification study by Ishikawa (2003), the observed temperature curves of all sensors at NAR indicate a high probability for the occurrence of permafrost, but require further data to confirm this. The lowering of MAGST values throughout the investigated time span is attributed to lower air temperatures in the 2015/2016 period compared to the 2014/2015 period, as recorded at the weather station in Samedan (MeteoSchweiz, 2018). The simultaneously recorded decrease in subsurface temperatures values shows a close coupling between the subsurface and the atmosphere. The observation of higher winter temperatures at positions where the sensors are placed at concave surface positions is attributed to a relatively thick snow cover that prevents the penetration of cold winter temperatures.

Longitudinal ERI models of the upper rock glacier lobes (E02/E03, E13, E15) display highest ice contents at positions below the talus cones. Sharp drops in th e longitudinal resistivity gradients indicate strong reductions of ground ice towards the adjacent rock glaciers. A qualitatively similar drop in resistivity at Schiantala rock glacier in the Italian Alps (Ribolini et al., 2010) was interpreted as the margin of a body buried ice from a LIA glaciation. If this assumption can be transferred to NAR, however, is speculative and not backed by any further observations. Transverse ERI models of the upper rock glacier lobes show relatively high ice contents at the lateral margins of the rock glaciers (E04: N; E10: B, C1; E11: C2; E12: C3), while the transitional zones between the rock glacier lobes are ice-free. This distribution of ground ice coincides with variations in rockfall intensity, as the transitional zones between the rock glaciers are less affected. It further coincides with MAGST values, which are higher at the ice-free positions in the transitional areas. However, at the transition between the rock glaciers NAR02 and NAR07, the observed ground ice distribution pattern is reversed. This is attributed to the extraordinary position of NAR07, which superimposes the joint root zone of NAR03 and NAR05. Presented ERI models of the lower rock glacier lobes show mainly

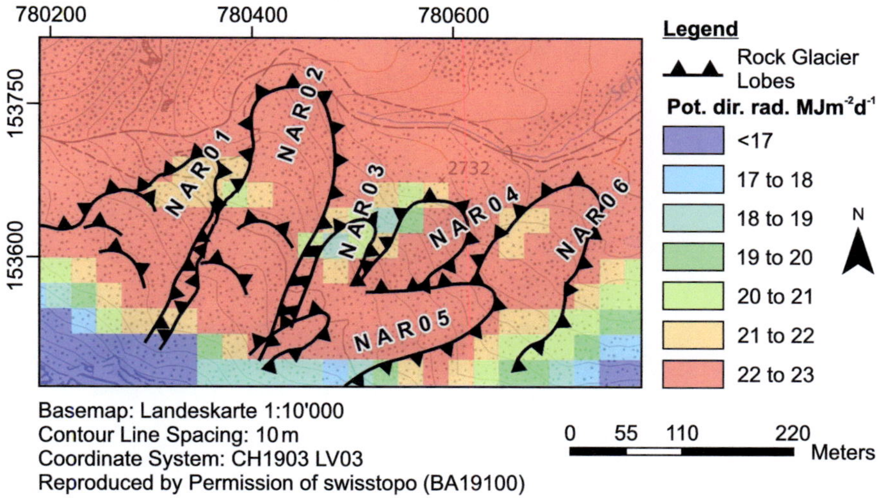

Figure 4.15: Map of Potential Direct Solar Radiation in Study Area NAR between July and October. Own Calculations. Source of DEM: swisstopo (2005).

unfrozen conditions (E). Extremely low resistivity values in the models E17 and E21 indicate accumulations of liquid water at the rock glacier snouts of NAR02 (E17: B) and NAR05 (E21: K). At NAR02, this reservoir presumably supplies the small spring at the foot of the rock glacier. However, isolated patches of ground ice were detected in the lower parts of the rock glaciers NAR04 and NAR06 (E20: H; E24: M, N).

The observed permafrost distribution at NAR correlates with the potential direct solar radiation between July and October (fig. 4.15), as calculated by the *Area Solar Radiation*-tool of ArcGIS (Ver. 10.6, ESRI Inc.) from a digital elevation model with a raster size of 25 m (swisstopo, 2005). Following the approach of Funk and Hoelzle (1992) and Hoelzle (1992), values of this parameter can be transferred (see tab. 4.2) into values of the Bottom Temperature of Snow Cover (BTS) (Haeberli, 1973) and thereby used as an indicator of permafrost. The presented map (fig. 4.15) shows that the position of the ground ice occurrence at NAR04 (E20: H) receives radiation values between 18.8 MJ m^{-2} d^{-1} and 19.3 MJ m^{-2} d^{-1}. This corresponds to BTS values below −3 °C, which clearly indicate permafrost conditions. Radiation values further reflect the absence of permafrost at the rock glacier snout of NAR05. However, the presented map fails in reflecting the assumed absence of permafrost at NAR03, where relatively low radiation values disagree with the relatively low resistivity values of the E19 model (E19: G). It further disagrees with the observation of ground ice at NAR06, which was verified by core drilling. At the corresponding position, radiation values are between 21.4 MJ m^{-2} d^{-1} and 21.6 MJ m^{-2} d^{-1}, which indicates the absence of permafrost, following Haeberli (1973). These discrepancies may be explained by the general uncertainty of the BTS method (Haeberli, 1973), but may also reflect a temporal disparity between the recent environmental conditions and the internal structure of the rock glaciers at NAR. This emphasizes the ability of the internal structure to serve as an archive for landforming processes and conditions of the past.

Chapter 5

Results and Interpretation UER

Geophysical investigations at UER comprise multi-dimensional approaches of ERI (5.1) and GPR (5.2). Temperature data was recorded at the ground surface and by a subsurface thermistor chain (5.3). The last part of this chapter comprises an interpretation of the results, focusing the internal structure of UER01 rock glacier (5.4).

5.1 Electrical Resistivity Imaging

The section is divided into four parts: The first three parts comprise results of the q-3-D ERI surveys (E01, E2, E03) at UER01. In the fourth part of this section, results of the 2-D ERI reconnaissance survey (E04) at UER02 are presented. This last part of the section includes an instant interpretation of the results, as no complementary data from UER02 is available. The more comprehensive interpretation of the internal structure of UER01 rock glacier is presented in section 5.4.

Results of a previous study, comparing results of 2-D ERI surveying, 2-D SRT surveying and borehole temperature measurements (Emmert and Kneisel, 2017), show that a value of around 8 kΩm can be used to distinguish between frozen and unfrozen subsurface conditions at UER. This threshold value is hence used for interpretations in this chapter. Although results of the surveys E01 and E02 have already been published (Emmert and Kneisel, 2017), they are rearranged here for a holistic view on the entire rock glacier.

UER01 Rock Glacier Snout

The q-3-D ERI model E01 (fig. 5.1) shows a complex resistivity distribution pattern. Resistivity values that exceed the value of 8 kΩm, cluster in three prominent, longitudinally shaped structures (A1–A3). Values of up to 14 kΩm (A1), 30 kΩm (A2) and 40 kΩm (A3), respectively, indicate frozen, but ice-poor conditions. The three structures root in a joint root zone (B), have lengths between 60 m and 70 m and widths of around 10 m. Their thicknesses and depths vary strongly between and within the extent of the three structures. The boundaries between the structures vanish below a depth of 6 m, but resistivity values are still above 12 kΩm at all three positions throughout the entire model depth of 15 m. Values of low resistivity are present at the western margin of the model, where a curved band of relatively low resistivity values ($<$1 kΩm) appears below a depth of 2 m (C). Its position corresponds to a surface ridge of around 2 m that forms the outer edge of the rock glacier in this part. A second structure of relatively low resistivity values appears as a U-shaped structure (D) in front of structure A2. At this position, which corresponds to a deep furrow between the arcuate ridges, values are below 2 kΩm through the entire model depth.

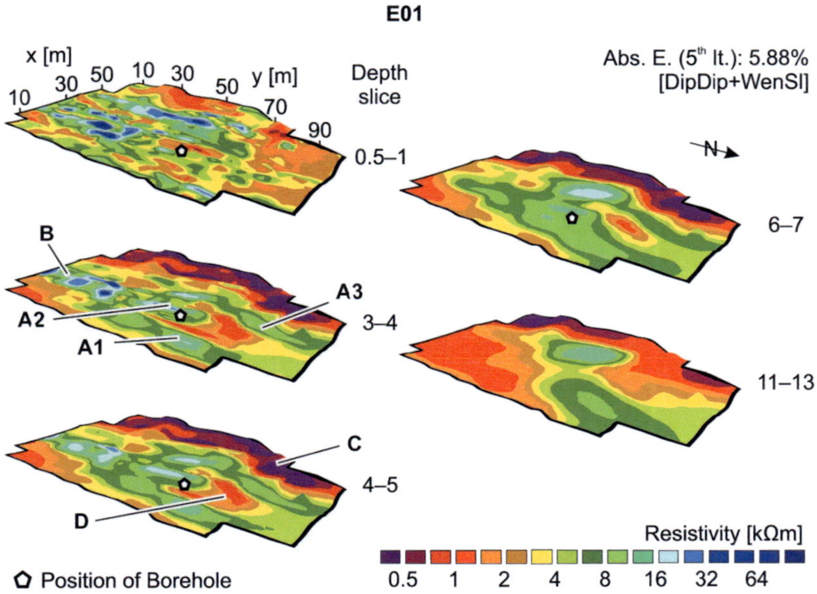

Figure 5.1: ERI Model UER: E01. Selected Slices. Labels Refer to Structures Mentioned in the Text.

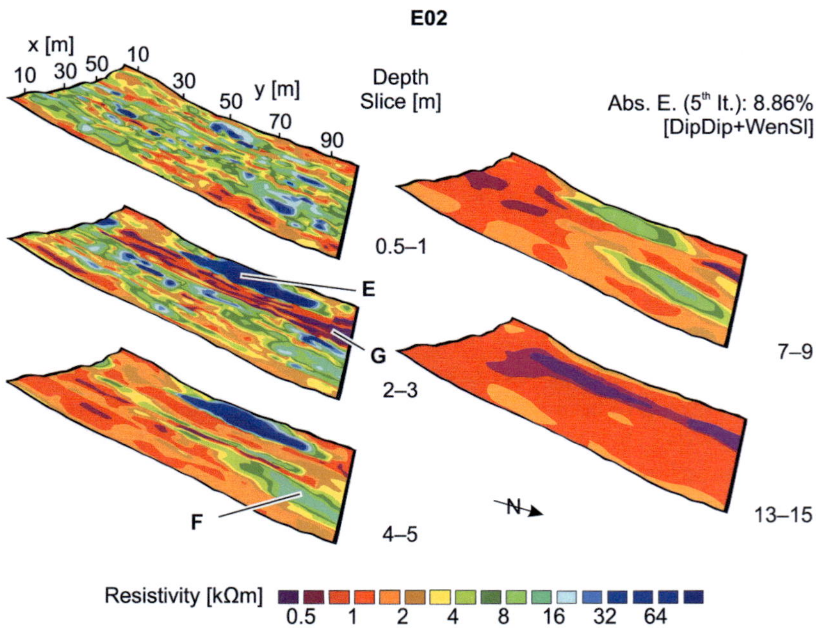

Figure 5.2: ERI Model UER: E02. Selected Slices. Labels Refer to Structures Mentioned in the Text.

UER01 Central Part

The uppermost slice of the E02 q-3-D ERI model, which represents the central part of the rock glacier, shows a complex resistivity distribution pattern (fig. 5.2). Except for a few isolated clusters of relatively high resistivity values in the upper subsurface, resistivity values exceed the threshold value of 8 kΩm only at two spots (E, F). In the western part of the model, maximum resistivity values of up to 250 kΩm cluster in an elliptically-shaped structure and indicate ice-rich conditions (E). Parts of this structure appear directly below the surface and at a depth of 2 m, the structure reaches its maximum horizontal extent. The structure remains at the same position down to a depth of 5 m and is followed by a decrease in extent and resistivity. The structure disappears below a depth of 11 m, and the horizontal model slices show a rather uniform resistivity distribution pattern.

A second structure of relatively high resistivity values between 8 kΩm and 18 kΩm appears in the central part of the model (F). At the most downslope part of the E02 model, where it partially overlaps with the E01 model, structure F attaches to the joint root zone of the tongue-shaped structures of the E01 model (B). Below a depth of 4 m, the extent of structure F decreases with depth. Like structure E, it completely disappears below a depth of around 11 m. Very low resistivity values ($<$1 kΩm) form a longitudinal band with a thickness of about 12 m (G) that appears between depths of 1 m and 4 m between the two structures E and F. Structure G disappears below a depth of 4 m but reappears at the bottom of the model, at depths between 11 m and 15 m.

UER01 Rock Glacier Root Zone

The E03 q-3-D ERI model shows relatively high resistivity values in the central part of the root zone (H), enclosed by two longitudinal bands of relatively low resistivity values (fig. 5.3). These bands are located at the western and at the eastern margin of the ERI model. Strong contrasts in resistivity characterize the horizontal slices.

Maximum values of up to 150 kΩm, which indicate ice-rich conditions, appear in the upper 2 m of the model. With increasing depth, isolated patches of relatively high resistivity values merge to an L-shaped structure of resistivity values between 50 kΩm and 100 kΩm (J). The maximum horizontal extent of this structure is reached between 3 m and 5 m depth. It covers an area that corresponds to the surroundings of the ice patch on the rock glacier surface (J1), and to a longitudinal gully in the western part of the rock glacier (J2), respectively. Below a depth of 5 m, resistivity values of structure J decrease with increasing depth. The magnitude of this decrease is stronger in the longitudinal segment J1 than it is in the transverse segment of the structure J2. At a depth of 9 m, the longitudinal part of the structure vanishes and only one ellipsoid patch of relatively high resistivity values is left. Although the slight decrease in resistivity continues with increasing depth, values at the bottom of the model at about 20 m depth (not shown) are still above 50 kΩm.

UER02 Reconnaissance Survey

The first part of the E04 model corresponds to the area in front of the rock glacier and shows resistivity values between 1.5 kΩm and 7 kΩm throughout the entire model depth (fig. 5.4). This indicates unfrozen conditions with a varying content of liquid water. Lowest resistivity

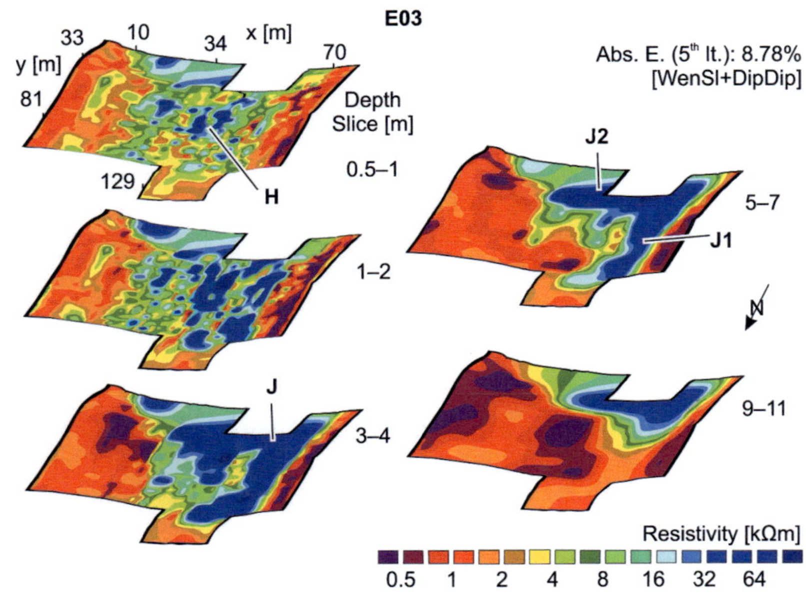

Figure 5.3: ERI Model UER: E03. Selected Slices. Labels Refer to Structures Mentioned in the Text.

values, between 0.2 kΩm and 1 kΩm, appear in a layer with a thickness between 1 m and 3.5 m (K). This structure indicates an interbedded layer of relatively fine-grain materials.

At a depth of around 4.5 m, a layer with relatively high resistivity values between 30 kΩm and 125 kΩm appears at the bottom of the model (L). It indicates frozen conditions and a high ice content. It can either represent ice-supersaturated conditions, like they are assumed for UER01, or massive ice, which is exposed about 100 m away from the survey location. However, further assumptions are speculative without complementary data. The undulating frost table topography follows the undulating surface topography. At the position where the survey line intersects with the upper lobe of the rock glacier, resistivity values of the upper layer decrease and values between 0.3 kΩm and 1 kΩm dominate the model.

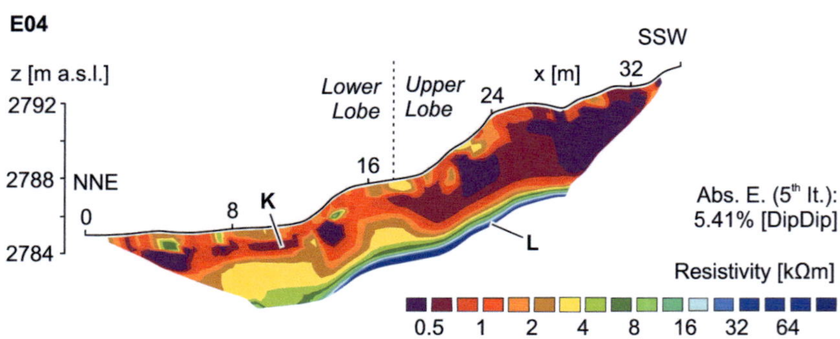

Figure 5.4: ERI Model UER: E04. Labels Refer to Structures Mentioned in the Text.

5.2 Ground-Penetrating Radar

One 2-D GPR survey was performed along the longitudinal axis of the rock glacier (R01) and three 2-D GPR surveys were performed in transverse direction: one across the root zone of the rock glacier (R02), one across the central part (R03) and one across the rock glacier snout (R04). For time-depth conversion, a simple two-layered 1-D model, based on hyperbola matching was used: it consists of an upper layer of 1 m thickness an EM velocity of 0.09 m ns^{-1}, which overlays a lower layer of an EM velocity of 1.2 m ns^{-1}.

Longitudinal Profile

Radargram R01 (fig. 5.5) shows a pattern of surface-parallel reflectors throughout the first 155 m (A). This pattern reaches down to a depth of around 10 m at the start of the survey line, but its thickness is reduced to around 6 m at $x = 50$ m. At several positions, the pattern of surface-parallel reflectors is interrupted by upwardly dipping reflectors at depths of around 6 m (B). Between $x = 165$ m and $x = 190$ m, a rather chaotic pattern of short reflectors appears (C). This pattern is followed in a downslope direction by surface-parallel reflectors (D), which are partially interspersed with nearly horizontally striking reflectors (E). In the part of the radargram that corresponds to the area where a distinct furrow-and-ridge topography is present, the shape of the reflectors follows the surface topography (F). Below the surface ridges, penetration depth is reduced from around 8.5 m to 6.5 m.

Transverse Profiles

The radargrams R02, R03 and R04 correspond to transverse cross-sections of UER01. The radargram of the root zone (R02:, fig. 5.6a) shows a low penetration depth which gradually increases from 2.5 m to 8 m in the first part of the radargram. This part corresponds to the area west of the surface ice patch and shows surface-parallel reflectors (G). In the part that corresponds to the area around surface ice patch it shows strictly horizontal reflectors. After the end of the surface ice patch, the pattern of horizontally striking reflectors continues for around 3 m and is covered by a shallow pattern of surface-parallel reflectors (J). In the part with inclining surface topography at the end of the radargram, it repeats the pattern of surface-parallel reflectors of the first 8 m and shows again a low penetration depth (K).

The R03 radargram, which corresponds to the central part of the rock glacier (fig. 5.6b) shows surface-parallel reflectors and a low penetration depth at the beginning of the survey line (L). In survey direction, this pattern is followed by a chaotic pattern of rather short reflectors (M) at positions that correspond to a surface depression. In a part that corresponds to the area of the longitudinal surface ridges, the radargram shows surface-parallel reflectors in the upper 2.5 m of the subsurface (N). Penetration depth is highest in the central part of the radargram but the differences between the identified sections are less pronounced compared to the R02 radargram. Horizontally striking reflectors (P) appear in the first part of the R04 radargram (fig. 5.6c). This pattern is covered with curved reflectors in the following, which appear directly below the surface (Q). While penetration depth is highest at positions that correspond to the furrows in the first half of the radargram (R), this pattern is reversed in the second half, in which penetration depth is lower at a similar positions (S).

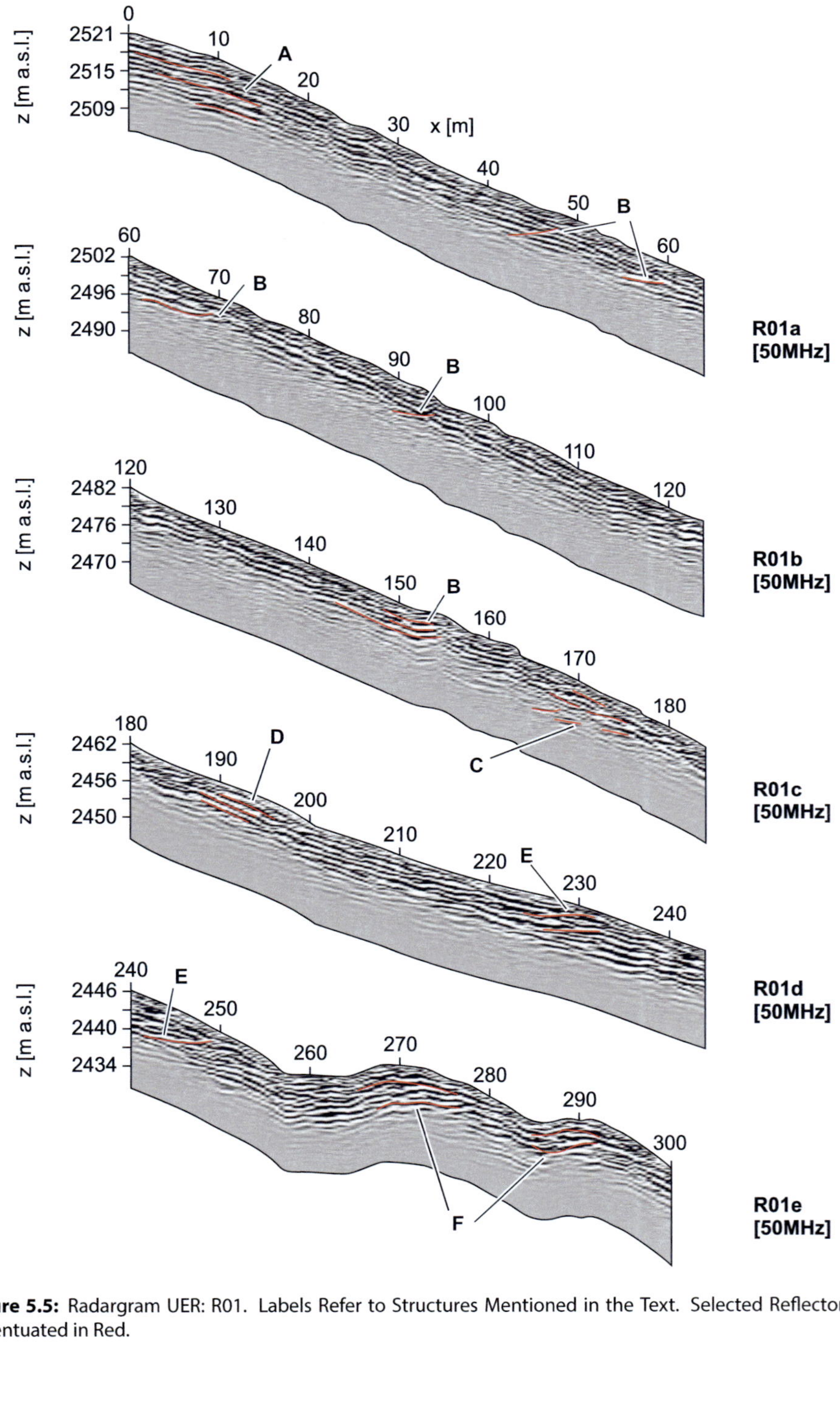

Figure 5.5: Radargram UER: R01. Labels Refer to Structures Mentioned in the Text. Selected Reflectors are Accentuated in Red.

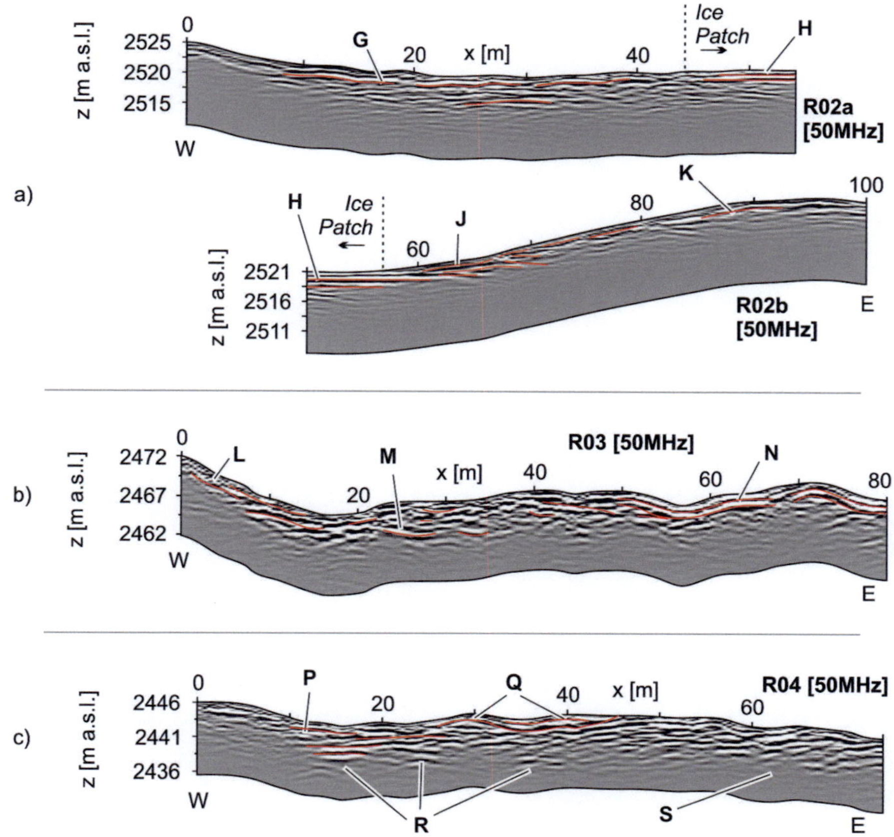

Figure 5.6: Radargrams UER: a) R02 b) R03 c) R04. Labels Refer to Structures Mentioned in the Text. Selected Reflectors are Accentuated in Red.

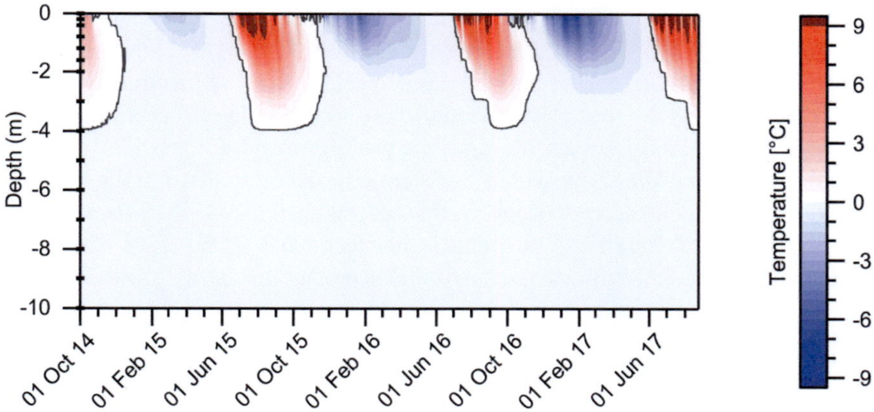

Figure 5.7: Recorded Borehole Temperature Data (UER). Small Dots on y-Axis Indicate Sensor Positions.

5.3 Temperature Measurements

Temperature data was continuously recorded at 15 temperature sensors, installed in a borehole of 10 m depth and at ten positions on the surface.

Subsurface Temperatures

Throughout the complete investigated period, subsurface temperature values at UER01 were below 0 °C at all sensors from a depth of 4 m. However, at all of these sensors, the difference between the highest temperature and 0 °C is below the accuracy range of the sensors (±0.25 °C). The borehole temperature record (fig. 5.7) shows that downward freezing from the surface started on November 16, 2014 in the winter of 2014/2015 (2015/2016: Oct 30, 2016/2017: Oct. 19), 19 days before the onset of upward freezing (2015/2016: 15 d, 2016/2017: 15 d). After 14 days (2015/2016: 15 d, 2016/2017: 22 d) the subsurface column was completely frozen and remained in a froze state for 175 days during the winter of 2014/2015 (2015/2016: 216 d, 2016/2017: 187 d). Thawing of the subsurface column took 37 days in the following spring of 2015 (2016: 58 d, 2017: 65 d) and reached a maximum thickness of 4 m on July 19, 2015 (2016: Aug 29, 2017: Aug 04). The recorded subsurface temperature values show an increasing strength in the penetration of cold winter temperatures throughout the investigated time span. The wintry cooling of 2014/2015 reached down to a depth of 1.6 m only, where it caused a temperature decrease of 0.31 K. The following winter of 2015/2016 caused a decrease of 0.65 K at the sensor in 2 m depth and in the winter of 2016/2017, the same sensor depth was affected by a temperature decrease of 1.19 K.

Ground Surface Temperatures

The temperature records of the ten GST sensors show a range of MAGST values between 3.23 °C (T03: 2014/2015) and −0.23 °C (T02: 2015/2016) (tab. 5.1). MAGST values of the second investigation period are generally lower than MAGST values of the first period. The difference between the two years is highest at the sensor locations T02 and T06, which represent the positions with lowest MAGST values. The lowest difference is observed at the sensors T05 and T07, which represent locations of relatively high MAGST values.

A comparison of MAGST values and the sensor positions reveals a connection between these two parameters: At the rock glacier snout, where arcuate ridge structures occur on the rock glacier surface, MAGST values are higher at positions on the ridges (T01, T03, T05), compared to the furrow (T02). This pattern is reversed in the central part of the rock glacier, where longitudinal ridge structures occur. In this central part, MAGST values are lower on the ridges (T06, T04), compared to positions in between (T07, T08). Plots of daily mean temperature values (fig. 5.8) indicate a strong influence of an insulating snow cover. At positions where a durable snow cover is assumed due to low diurnal temperature fluctuations (T02, T04 and T07), lowest daily mean temperature values are at or below −5 °C. Minimum temperatures at the other sensors are between −14.66 °C (T01) and −7.06 °C (T03). The order which sensor became snow-free first varied between the different sensors on the arcuate ridges (2015: T01, 2016: T01, 2017: T05), but the sensor at the deep surface depression (T02) was always the last one that became snow-free. The sensor at the apparently relict lobe (T09)

Table 5.1: Summary of Recorded GST Data (UER).

Label	MAGST [°C] 2014/2015[a]	MAGST [°C] 2015/2016[a]	SC [d] 2014/2015[a]	SC [d] 2015/2016[a]	SC [d] 2016/2017[b]	Topographic Position
T01	2.04	0.98	173	135	92	Arcuate Ridge
T02	1.85	−0.23	258	262	224	Bottom of Furrow
T03	3.23	1.49	232	190	190	Arcuate Ridge
T04	1.99	0.89	234	266	204	Longitudinal Ridge
T05	2.07	1.17	246	178	160	Arcuate Ridge
T06	2.17	0.13	236	177	158	Longitudinal Ridge
T07	2	1.36	259	261	241	Between Ridges
T08	2.59	0.79	226	171	165	Between Ridges
T09	1.76	0.02	257	202	172	Relict Lobe
T10	−0.09	−1.56	234	208	170	Scree Slope

MAGST = Mean Annual Ground Surface Temperature // SC = Snow Cover
[a]Investigated Period: 01 Oct–30 Sep // [b]Investigated Period: 01 Oct–25 Aug

shows a similar curve compared to the other sensors. Temperature values are relatively low in winter and moderate in summer. The sensor at the adjacent scree slope (T10), however, shows significantly lower MAGST values compared to the sensors at the rock glacier surface. This is attributed to the absence of an insulating snow cover, which causes extraordinary low winter temperatures at this position.

5.4 Interpretation

The observation of relatively low MAGST values at the adjacent scree slope (T10) and the apparently relict lobe in front of the rock glacier (T09), indicate that the occurrence of permafrost conditions is not restricted to the recent outline of UER01. However, highly fluctuating temperature values indicate the absence of an insulating snow cover at these positions. Therefore, assumptions on the presence of permafrost cannot be made from temperature records alone (Ishikawa, 2003). This ambiguity is reflected in the conflicting representation of the investigated area by maps of the potential permafrost distribution (BAFU, 2005; Boeckli et al., 2012). The observed differences in MAGST values between the two investigated periods follow interannual variations in air temperatures, as recorded at the weather station in Samedan (MeteoSchweiz, 2018). The enhanced subsurface cooling is attributed to decreases in snow cover thickness and duration. This is indicated by (i) a shortening of the time span with low diurnal fluctuations and (ii) a shortening of the zero-curtain period in the GST records. The assumption is further supported by snow height measurements from the nearby Piz Kesch/Porta d'Es-cha station (MeteoSchweiz, 2018).

The surface depression in the central western part of the rock glacier has presumably the highest ice content. The relatively high resistivity values (E02: E) at this position, however, coincide with a chaotic pattern of short reflectors in the transverse R03 radargram (R03: M). This indicates a large amount of debris in the subsurface and the absence of massive ground ice (Berthling et al., 2000; Degenhardt, 2009). A frozen, ice-saturated state of the subsurface materials is further indicated by the relatively deep penetration depth of the EM waves at this

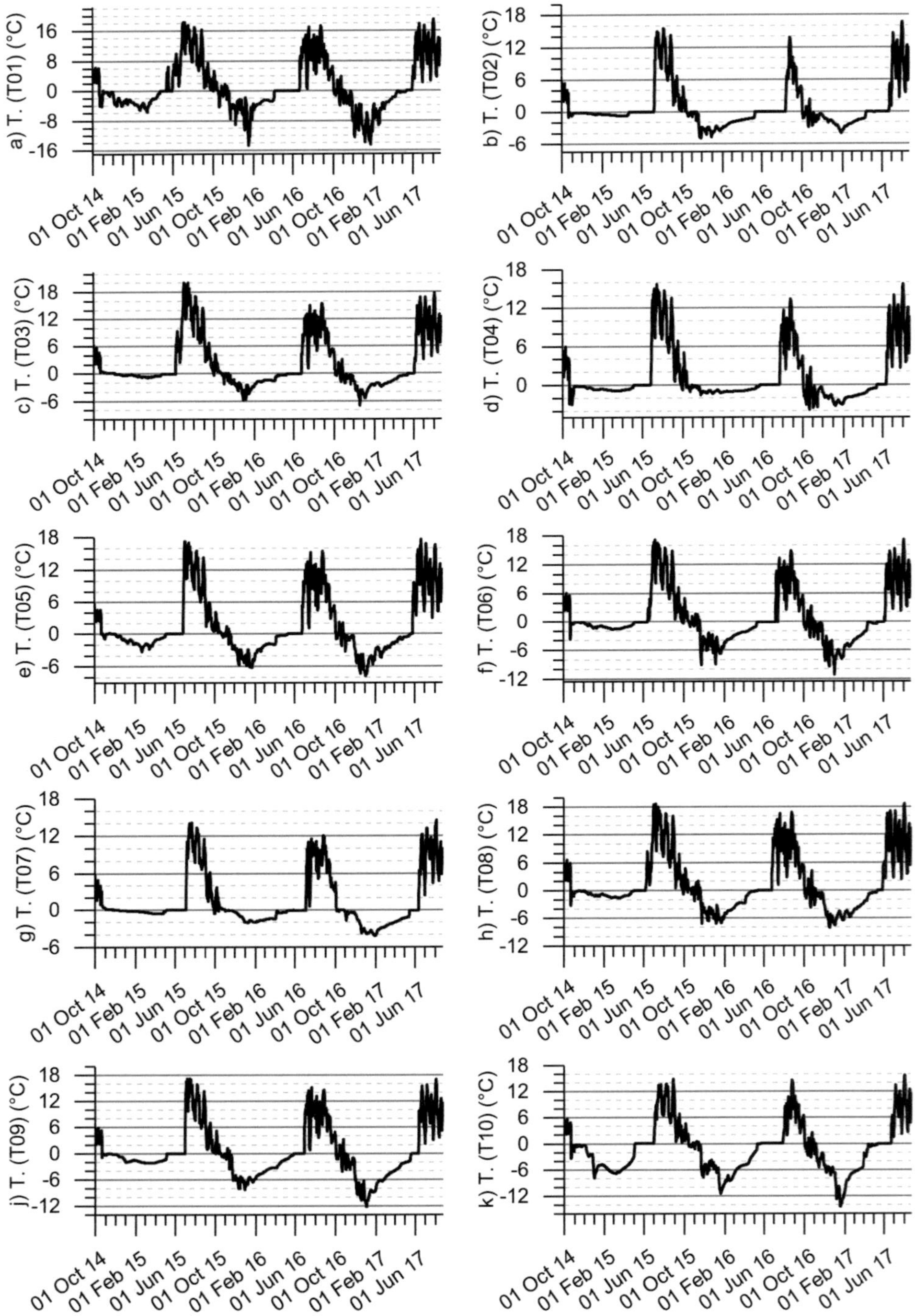

Figure 5.8: Recorded GST Data UER.

position (Scott et al., 1990). A durable snow cover was observed in the surface depression during fieldwork, and thus it is assumed that either (i) meltwater percolates into the subsurface and refreezes, or that (ii) snow is buried by material from the rim of the depression.

The second area with presumably high ice content is located in the root zone of UER01, directly attached to the surface ice patch (E03: J). Resistivity values are close to 100 kΩm and, regarding the local conditions, are assumed to indicate congelation ice. Although a lateral moraine and the perennial surface ice patch indicate a glacial influence on the formation of permafrost, the R01 radargram (R01: A) lacks any indications of massive ice, like e.g., blank areas or synclinal structures (Monnier et al., 2011; Monnier et al., 2013; Moorman et al., 2003). Hence, it is unlikely that the ground ice is an extension of the surface ice patch, which reaches down to a depth of less than 2 m, as indicated in the R02 radargram (R02: H).

Similar reflectors, as they appear in the R01 radargram (R01: A), were observed by Degenhardt (2009), regarding rock glaciers in the USA and Svalbard, and interpreted as a result of sequential covering of surface ice by debris layers. This concept is also plausible for UER01 rock glacier, as the surface ice patch is located in a local depression of the rock glacier surface and therefore might be covered by debris from time to time. The longitudinal part of the ground ice body at the western side of the rock glacier (E03: J1) is attributed to refreezing meltwater of the surface ice patch. Upwardly dipping reflectors in the R01 radargram (R01: B) are commonly attributed to thrusting processes (Monnier et al., 2011). Their appearance in a part of UER01 where longitudinal ridges occur, indicates that these surface features result from permafrost creep along the longitudinal axis of the rock glacier.

At the rock glacier snout, where the distinct furrow-and-ridge topography occurs, the surface-parallel reflectors of the R01 radargram (R01: F) show an impact of strong compressional forces (Degenhardt and Giardino, 2003; Monnier et al., 2008). Small-scale differences in penetration depth and reflection amplitude in the transverse R04 radargram of the rock glacier snout (R04: R, S) affirm the occurrence of a complex ground ice distribution pattern, as derived from the E01 models. In the first part of the radargram, the observation of a low penetration depth at positions that correspond to the arcuate ridges, however, contrasts their assumed frozen state (Moorman et al., 2003). This discrepancy is associated with the observation of vanishing boundaries between the ground ice patches at greater depth levels.

Although no velocity measurements were performed at UER01, the patchy ground ice distribution and the lack of an extensive frost table at UER01 point towards an inactive state (Barsch, 1996). This is affirmed by the patchy occurrence of pioneer plants, but due to the particularly small grain size, conventional concepts that link vegetation and landform activity may not be transferable to pebbly rock glaciers (Burga et al., 2004).

Chapter 6

Results and Interpretation LTF

Geophysical investigations at LTF comprise one q-3-D ERI survey (6.1) and multi-dimensional approaches of GPR (6.2). The last part of this chapter comprises an interpretation of the results, focusing the internal structure of the investigated rock glacier (6.3).

6.1 Electrical Resistivity Imaging

A previous study from LTF rock glacier (Ikeda and Matsuoka, 2006), which included an overlay comparison between a direct observation of ground ice in the fine-grained section of the survey area and results of 2-D ERI surveying, showed that frozen conditions can be present in areas with resistivity values above 1 kΩm. The horizontal resistivity distribution of the q-3-D ERI model E01 (fig. 6.1) reflects the segmentation of the survey area into two main sections, covered with clasts of different origin and grain size. While the resistivity distribution of the western section, where fine-grained shale debris is exposed, is rather uniform and dominated by relatively low resistivity values, the eastern section, which is covered by coarse-grained dolomite debris, shows a more complex distribution pattern. The eastern part is further subdivided into an upward, northeastern part dominated by relatively low resistivity values and a downward, southeastern part that shows relatively high resistivity values. This segmentation is present throughout the entire vertical model range of 15 m. Resistivity values are between around 128 kΩm at isolated patches in the northeastern corner and around 1 kΩm at patches at the southeastern margin of the model. A transitional area exists between these two extremes.

Around the northeastern corner of the model (A), resistivity values of around 25 kΩm appear directly below the surface. This location corresponds to the area close to the front of the upper rock glacier lobe in the coarse-grained section. Resistivity values decrease with increasing depth and reach around 10 kΩm at 5 m depth. With further increasing depth, resistivity value increase to around 30 kΩm at a depth of 9 m. The horizontal extent of this structure of relatively high resistivity values changes only slightly throughout the entire model depth. At the opposite corner of the model (B), resistivity values are mainly below 2 kΩm. This part of the model lacks a distinct vertical layering and shows only a slight increase from around 1 kΩm to 2 kΩm within the upper 5 m of model depth. Below this depth, resistivity values vary only slightly around 2 kΩm. At the position of the borehole of Ikeda and Matsuoka (2006), relatively low resistivity values of around 1 kΩm are present in the upper 5 m of the subsurface. Below this depth, resistivity values increase to 4 kΩm at 10 m depth and to more than 5 kΩm at the bottom of the model.

The central part of the model (C), which corresponds to the area around the front of the upper rock glacier lobe in the fine-grained section, shows constantly increasing resistivity

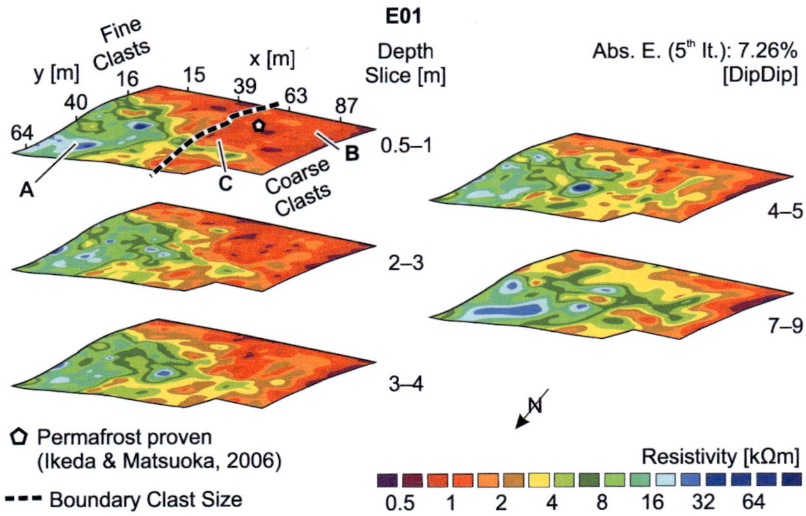

Figure 6.1: ERI Model LTF: E01. Selected Slices. Labels Refer to Structures Mentioned in the Text.

values with increasing depth. This part of the model represents a transitional zone between the two extremes. The lateral extent of this zone increases with depth and it has a distinct vertical layering: an upper layer of resistivity values around 2 kΩm and 3.5 m thickness covers a lower layer with resistivity values of around 6 kΩm. This lower layer reaches down to a depth of around 8 m. Below this depth, resistivity values decrease and are around 1 kΩm at the bottom of the model.

6.2 Ground-Penetrating Radar

Two 1-D CMP surveys and four 2-D GPR surveys were performed at LTF. Two one-dimensional models of the vertical EM velocity layering were derived from the two 1-D CMP surveys, each model representing one of the two main sections of the rock glacier. These models were then used for a time-depth conversion of the data from the longitudinal surveys of the corresponding rock glacier section. More complex 2-D models of the vertical EM velocity layering were built by combining the two 1-D models of vertical EM velocity layering. These models were used for a time-depth conversion of the data from the transverse surveys.

CMP Surveys

The R01 CMP model (fig. 6.2a) shows a vertical EM velocity layering that consist of an upper layer with a thickness of 0.8 m and an EM velocity of 0.12 m ns^{-1}. It is followed by a layer with a thickness of 2.3 m and an EM velocity of 0.09 m ns^{-1}. Below a depth of 3.1 m, a layer with an EM velocity of 0.14 m ns^{-1} appears. The R02 CMP model (fig. 6.2b) shows a similar layering, but with slightly different depths of the boundaries between the layers and a slightly different EM velocity of the upper layer. The upper layer has an EM velocity of 0.11 m ns^{-1}

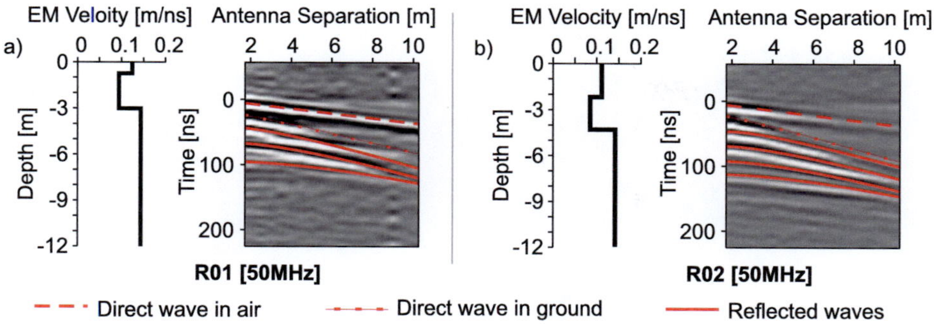

Figure 6.2: Radargrams LTF (Right Panels) and Derived Velocity Models (Left Panels). a) R01, b) R02. Selected Reflectors are Accentuated in Red.

and reaches down to a depth of 2.2 m. It is followed by a layer with 2.1 m thickness and an EM velocity of 0.09 m ns^{-1}. Below a depth of 4.3 m, the EM velocity is 1.4 m ns^{-1}.

Longitudinal Surveys

The R03 radargram (fig. 6.3a) corresponds to a longitudinal cross-section of the fine-grained section and shows surface-parallel reflectors within the upper 3 m of the radargram (A). This pattern can be traced throughout the complete radargram. One single, concave reflector, which reaches into a depth of around 5 m, represents the lower boundary of this pattern throughout the first 30 m of the R03 radargram (B). When the radargram corresponds to the lower lobe of the rock glacier, horizontally striking reflectors (C) appear below the surface-parallel reflectors and penetration depth reaches up to 6.5 m. Pattern C ends when the radargram corresponds to the front of the lower lobe, and penetration depth decreases significantly (D). Surface-parallel reflectors (E) appear within the first 30 m of the R04 radargram (fig. 6.3b), which corresponds to a longitudinal cross-section of the coarse-grained section. Penetration depth is low in this part of the radargram and reaches up to 5 m only. In the second part of the radargram, the orientation of the reflectors changes slightly into a rather horizontal direction and penetration depth increases to around 9 m (F).

Transverse Surveys

The part of the R05 radargram (fig. 6.4a) correspond to a cross-section of the fine-grained section of the upper rock glacier lobe. This part of the radargram shows a pattern of narrow, nearly surface-parallel orientated reflectors (G). It is covered by a shallow layer with around 1.5 m thickness, in which the reflection amplitude is relatively weak (H). In the second part of the radargram, which corresponds to the coarse-grained section, the separation between the reflectors of the surface-parallel pattern is more distinct (J). Penetration depth increases from around 3.5 m to around 5 m towards the second part of the radargram. The R06 radargram (fig. 6.4b) shows a pattern of narrow, surface-parallel reflectors in the upper 3 m of the subsurface (K). Below this depth, signal attenuation is high and only weak reflections are visible in deeper parts of the first part of the radargram. The second part of the radargram,

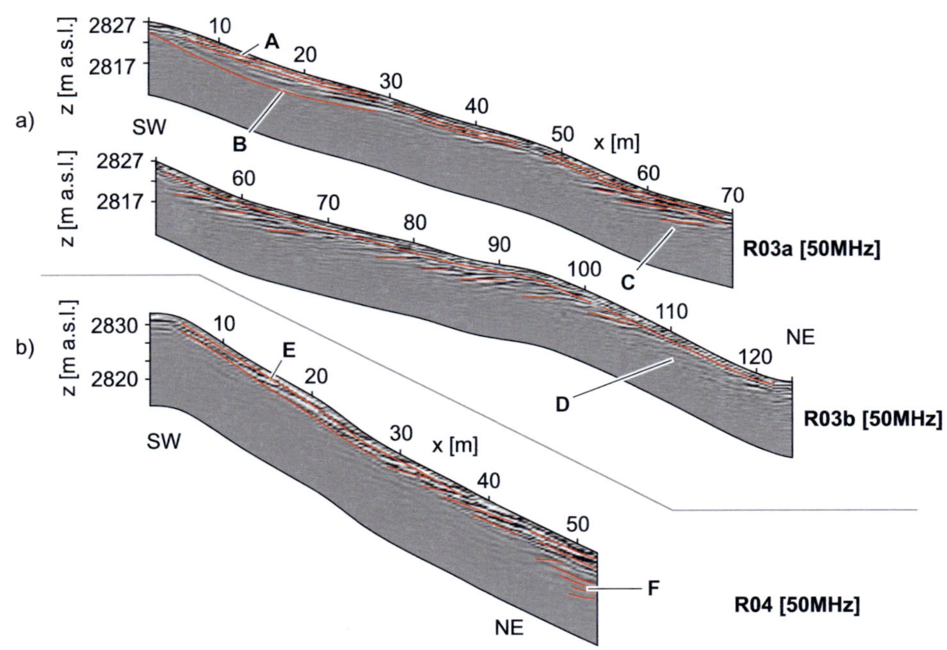

Figure 6.3: Radargrams LTF: a) R03 b) R04. Labels Refer to Structures Mentioned in the Text. Selected Reflectors are Accentuated in Red.

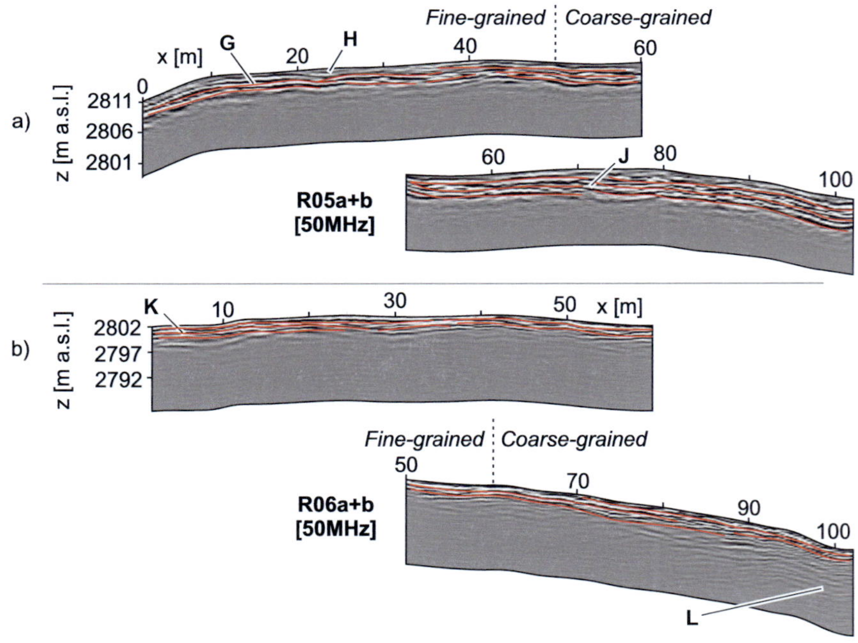

Figure 6.4: Radargrams LTF: a) R05 b) R06. Labels Refer to Structures Mentioned in the Text. Selected Reflectors are Accentuated in Red.

which corresponds to a cross-section of the coarse-grained section, shows a higher reflection amplitude and a higher penetration depth. A maximum penetration depth of about 12 m (L) reveals that the surface-parallel reflection pattern continues at greater depths (L).

6.3 Interpretation

When a resistivity value of 1 kΩm, which represents frozen materials (Ikeda and Matsuoka, 2006), is used as a lower boundary of permafrost conditions, the fine-grained section is classified as nearly completely frozen below depths of 3 m and 4 m. However, at the position of the borehole, where the occurrence of permafrost was proven in the past (Ikeda et al., 2008), modeled resistivity values exceed this threshold only below a depth of 5 m. Although values of single model cells at the borehole position exceed 1 kΩm already between depths of 0.5 m and 2 m, conclusions on a frozen state cannot be drawn here. This is because the exceed over the threshold value is below 5 %, which is the accepted level of deviation between two reciprocal measurement (see sect. 3.1). However, it must be noted that georeferencing of the horizontal resistivity slices is based on the retrieval of only eight color markers on the rock glacier surface, which are used as ground control points. The low number of ground control points may be responsible for geometrical inaccuracies that can lead to differences between the actual borehole location in the field and its assumed position in the ERI model.

Increasing EM velocity values from 0.09 m ns^{-1} to 0.14 m ns^{-1} at a depth of about 3.1 m in the R01 CMP model indicate a transition from an unfrozen to a frozen state (Moorman et al., 2003). Very low values directly above the assumed frost table indicate a high content of liquid water. This aligns with the description of a layer of "cobbles/pebbles with sandy/silty matrix" that was visually observed by Ikeda and Matsuoka (2006) directly above the frost table. A similar subsurface layering is observed by a comparison between the EM velocity layering of the R01 CMP model and a corresponding vertical resistivity plot of the ERI model (not shown). The R02 CMP model shows a similar increase in EM velocity values from 0.09 m ns^{-1} to 0.1 m ns^{-1} at a depth of about 4.3 m. When this depth is transferred to a corresponding position of the ERI model, it coincides with a vertical resistivity increase from 2 kΩm to 3.5 kΩm (not shown).

This observation indicates that a higher resistivity threshold can be used to distinguish between frozen and unfrozen conditions in the coarse-grained section of the survey area, compared to the fine-grained section. This difference can be explained by a lower water content or lithological differences (Etzelmüller et al., 2006; Schneider et al., 2013). The greater depth at which the frost table is apparently located in this coarse-grained section of the survey area is compensated partly by an ascending surface elevation from the fine-grained section towards the coarse-grained section. Therefore, the frost table is assumed to be at a similar elevation in both sections of the rock glacier, while the thickness of the active layer increases from the fine-grained section towards the coarse-grained section. This shows a higher volume of the debris accumulations on the rock glacier surface below the dolomite rockwall. This higher volume is reflected by the wider separation between the surface-parallel reflectors in the corresponding part of the transverse R05 radargram (R05: J).

The R03 radargram lacks a distinct boundary that could be interpreted as a frost table. This is attributed to the water-permeability of the frost table at LTF study site (Ikeda et al.,

2008), as this affects the physical properties of the subsurface layers and reduces the contrast between the specific dielectric constants (Berthling and Melvold, 2008). The occurrence of a water-permeable frost table and a high amount of liquid water in the frozen layer further agrees with the relatively low resistivity values that represent frozen materials. The appearance of one prominent concave reflector in the R03 radargram at a position close to the scarp (R03: B) can be interpreted as (i) the boundary between debris and bedrock or (ii) an internal bedding plane (Monnier et al., 2011; Otto and Sass, 2006). As very low resistivity values in the corresponding part of the ERI model are in a range that represents unconsolidated debris, the second hypothesis is preferred. The reflection patterns of the radargrams R03 and R04, which show surface-parallel (R03: A, R04: E) and horizontally striking reflectors (R03: C; R04: F), show characteristics of relatively undisturbed stratified talus deposits (Sass, 2006).

The position where highest resistivity values appear in the ERI model (E01: A) corresponds to positions of highest penetration depth in the radargrams R04 and R06 (R04: F; R06: L). Both observations indicate a relatively high ice content and align with the local accumulation of coarse clasts in this part of the rock glacier (Haeberli and Vonder Mühll, 1996; Moorman et al., 2003). The occurrence of accumulations of coarse clasts is often connected to a lowering of surface and subsurface temperatures, as, due to a higher density, warm air between the clasts is replaced by cold air (Harris and Pedersen, 1998; Hoelzle et al., 1999; Ishikawa, 2003; Kneisel et al., 2000; Rödder and Kneisel, 2012a). However, a characteristic permafrost layering that consist of an upper layer of relatively lower resistivity values and a lower layer of relatively higher resistivity values is absent in this part of the ERI model. Relatively high resistivity values appear down from the first model slice and decrease with increasing depth. Possible explanations are (i) an extremely shallow active layer that is not resolved in the ERI model or (ii) the existence of air-filled voids between the clasts (Vonder Mühll et al., 2002).

Chapter 7
Results and Interpretation FUR

Geophysical investigations at FUR comprise different approaches of 3-D ERI (7.1) and 2-D GPR (7.2). The last part of this chapter comprises an interpretation of the results, focusing the internal structure of the investigated lobes (7.3).

7.1 Electrical Resistivity Imaging

The q-3-D ERI model E01 (fig. 7.1) shows a complex resistivity distribution pattern that complicates a delimitation of discrete structures. Values range mainly between 0.1 kΩm and 50 kΩm, but single cells at the bottom of the model reach higher values of up to 100 kΩm (not shown). The horizontal resistivity distribution of the upper subsurface broadly reflects the observed surface morphology: highest resistivity values appear in the back of the lobe fronts and decrease in an upslope direction. However, this characteristic pattern disappears with increasing depth. Relatively high resistivity values cluster at two spots (A, B): At spot A, which corresponds to the position of a compound structure of several lobes, relatively high resistivity values appear down from the first model slice. Resistivity values of single cells within the upper 2 m of the model exceed 40 kΩm, but values between 5 kΩm and 10 kΩm dominate this part. Between 2 m and 3 m depth, resistivity values of the model slice are generally higher and in a range between 10 kΩm and 20 kΩm. However, also the extremely high resistivity values (>20 kΩm) at spot A disappear. Resistivity values further increase with depth and reach values of up to 40 kΩm at a depth of 7 m. This increase is followed by an abrupt drop to resistivity values between 4 kΩm and 7 kΩm.

A second spot of relatively high resistivity values appears in the central part of the model (B). This area corresponds to a part of the slope, which is only partly affected by the occurrence of lobes. Relatively high resistivity values are visible below a depth of 2 m, as an attachment to the cluster of relatively high resistivity values at spot A. The horizontal extent of the area of relatively high resistivity values increases with depth and reaches its maximum at a depth of 7 m. While the extent decreases with further increasing depth through the rest of the E01 model, the resistivity values at spot B increase from 10 kΩm between depths of 5 m and 7 m to 15 kΩm between depths of 7 m and 9 m. Maximum resistivity values of around 30 kΩm appear below a depth of 9 m. This cluster of relatively high resistivity is covered by a layer of extremely low values (<2 kΩm) in the shallow subsurface.

The E02 model (fig. 7.2a), which corresponds to the surrounding of the upper lobe FUR01, is dominated by relatively low resistivity values, mainly between 1.5 kΩm and 3 kΩm. Only single cells at the upper 1.5 m and below a depth of 3.5 m show values of up to 30 kΩm. The horizontal slices show that the resistivity values in the central part of the model (C) are lower than the resistivity values at the margins of the model (D, E). This distribution pattern re-

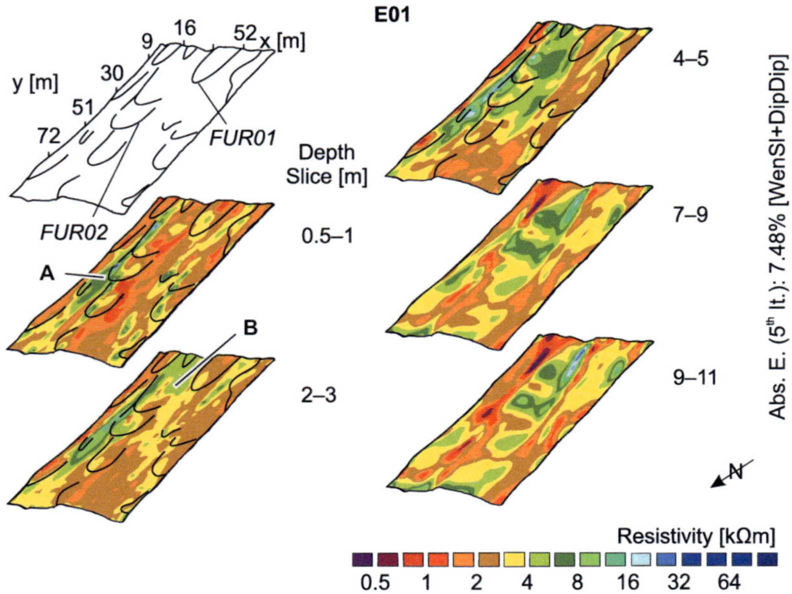

Figure 7.1: ERI Model FUR: E01. Selected Slices. Labels Refer to Structures Mentioned in the Text.

flects the outline of lobe FUR01: relatively high resistivity values appear at positions that correspond to the lateral risers of the lobe, while resistivity values at positions that correspond to the tread are relatively low. One exception is the appearance of a small isolated patch of relatively high resistivity values (up to 6 kΩm), at a position that corresponds to the central part of the lobe tread (F).

The E03 model (fig. 7.2b), which corresponds to the surrounding of lobe FUR02, shows generally higher resistivity values, compared to the E02 model. They are mainly between 2.5 kΩm and 6 kΩm but reach up to 50 kΩm in the upper 1.5 m of the subsurface and again below a depth of 5 m. Lowest resistivity values concentrate directly below the ground surface at a central position that corresponds to the lobe tread (G). This structure is enclosed by relatively high resistivity values at positions that correspond to the risers of the lobe. Resistivity values at the western lateral riser (H) and at an isolated spot at the central part of the lobe tread (J) are particularly high and reach up to 8 kΩm between depths of 2.5 m and 3.5 m and to 13 kΩm between depths of 3.5 m and 5 m.

7.2 Ground-Penetrating Radar

Highly variable subsurface conditions prevent a successful assessment of the vertical EM velocity layering by CMP surveying and, as suitable hyperbolas were absent in the radargrams, a constant EM velocity value of 0.12 m ns^{-1} was used for time-depth conversion. This value was chosen as an average of EM velocity values for coarse gneiss debris, which range between 0.09 m ns^{-1} and 0.14 m ns^{-1} (Sass, 2007).

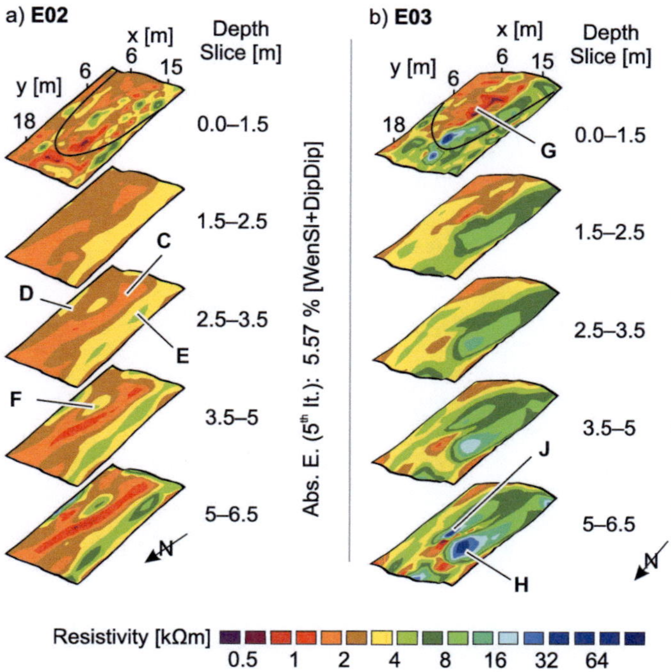

Figure 7.2: ERI Model FUR: a) E02, b) E03. Selected Slices. Labels Refer to Structures Mentioned in the Text.

The R01 radargram (fig. 7.3a) shows a relatively uniform subsurface layering, with a continuous pattern of surface-parallel reflectors in the shallow subsurface (A). Below a depth of around 3 m, this pattern is followed by weak reflections, which strike towards the inclining surface of the slope (B). The R02 radargram (fig. 7.3b) shows a similar pattern of surface-parallel reflectors in the shallow subsurface, compared to the R01 radargram. In contrast to the R01 radargram, however, this pattern is interrupted by multiple undulating reflectors (C). The appearance of these undulating reflectors is restricted to positions that correspond to lobe FUR02, and they are absent in other parts of the radargram. The low penetration depth in both radargrams R01 and R02 points to relatively moist conditions in the subsurface (Moorman et al., 2003).

7.3 Interpretation

The investigated lobes show a characteristic material sorting of lobate features on alpine talus slopes, like solifluction lobes and pebbly rock glaciers. This sorting includes the accumulation of rather coarse clasts at the frontal and the lateral risers, while rather fine-grained materials are present at the lobe treads (Benedict, 1970a; Matsuoka et al., 2005). This material distribution is reflected by the horizontal resistivity distribution of the ERI models, as the transport of electric current at the lobe risers is limited by higher amount of air-filled voids between coarse clasts (E02: D, E; E03: H) (Binley, 2015; Draebing and Eichel, 2017).

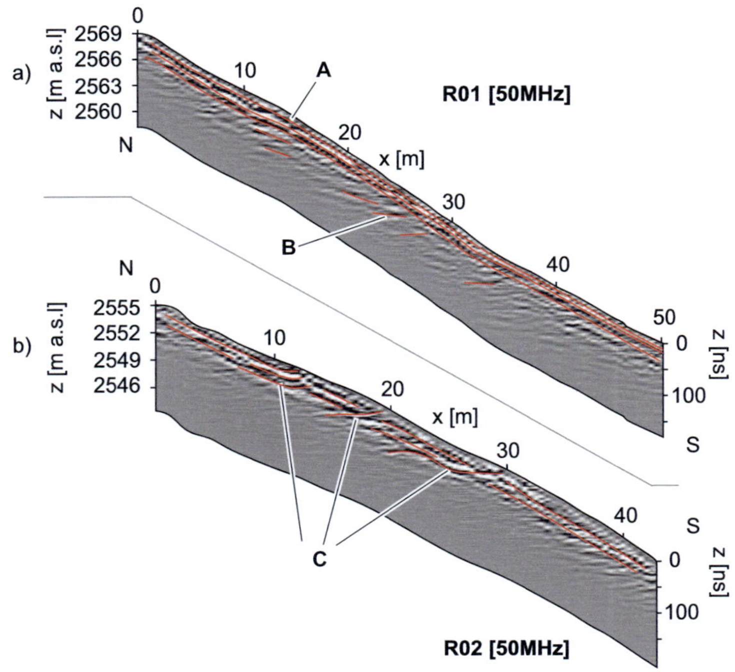

Figure 7.3: Radargrams FUR: a) R01, b) R02. Selected Slices. Labels Refer to Structures Mentioned in the Text. Selected Reflectors are Accentuated in Red.

Maximum resistivity values of the E01 model are in a range that represents frozen materials at talus slopes in comparable settings: At Les Attelas talus slope, Scapozza et al. (2011) derived resistivity values between 15 kΩm and 50 kΩm for frozen materials and Otto and Sass (2006) assumed that resistivity values above 20 kΩm represent frozen materials at a talus slope in Turtmanntal. As permafrost-favorable conditions for FUR study site are also depicted in maps of the potential permafrost distribution (BAFU, 2005; Boeckli et al., 2012), it is assumed that the two patches of relatively high resistivity values (E01: A, B) represent permafrost occurrences at different topographic positions and depth levels. According to Matsuoka (2001), the depth of movement can be derived from the height of a lobe. When this approach is transferred to the lobes at FUR study site (not shown), it shows that the depth of movement in the eastern part of the E01 survey area coincides within a cluster of relatively high resistivity values, representing frozen conditions (E01: A). It is consequently assumed that, as movement occurs within a frozen layer, this part of the slope is affected by permafrost creep. Thus, the lobes in this part of the slope (e.g., FUR01) are classified as pebbly rock glaciers (French, 2018; Matsuoka et al., 2005). In the northern part of the E01 survey area, the structure of relatively high resistivity values (E01: B) is located below the calculated depth of movement, and the lobes in this part are therefore classified as solifluction lobes.

This assumption is affirmed by the corresponding radargrams. The R01 radargram shows a characteristic reflection pattern of stratified talus deposits (Sass, 2006; Sass, 2007), while features that would indicate deformation processes are absent. Surface-parallel reflectors in the upper part of the R01 radargram (R01: A) indicates a uniform movement rate without vertical variations in velocity. The observed pattern is characteristic for solifluction lobes and was observed e.g., at solifluction lobes in Antarctica (Mori et al., 2007) and the Rocky Mountains (Leopold et al., 2008). The appearance of undulating reflectors in the R02 radargram (R02: C), however, is a recurring feature in radargrams of rock glaciers (e.g., Monnier et al., 2011) and indicates strong vertical variations in movement rates. The deformed shape of the reflectors resembles exemplary plots of movement rates from rock glaciers (Matsuoka et al., 2005) but not from solifluction lobes (Harris et al., 2008b; Matsuoka, 2001).

The occurrence of different lobe types on the talus slope is also indicated by the prominent differences in resistivity between the ERI models E02 and E03. The E02 model, which corresponds to lobe FUR01 is dominated by resistivity values presumably representing unfrozen conditions. Resistivity variations within the outline of the lobe (E02: D, E) are attributed to small-scale variation in material composition (Draebing and Eichel, 2017; Vanhala et al., 2009). The longitudinal band of relatively low resistivity values (E02: C) indicates that the position of FUR01 corresponds to a longitudinal channel in the subsurface with a relatively high content of liquid water. This assumption aligns with the observed low penetration depth in the R01 radargram (Moorman et al., 2003). However, it cannot be excluded that the strong signal attenuation is caused by the impingement of EM waves on the interface between debris and bedrock (Sass, 2007). Transferred to the R01 radargram, the height of FUR01 (around 1.5 m) corresponds to the depth of the transition between the surface-parallel reflectors (R01: A) and reflectors that strike towards the surface (R01: B). The lobe is hence assumed to comprise a deep-reaching solifluction component (Matsuoka, 2001).

When the height of FUR02 (around 1.8 m) is transferred to the E03 model, the resulting depth of movement coincides with a structure of relatively high resistivity values up to 13 kΩm (E02: H). Assuming that this range represents frozen materials, which is plausible following studies from other talus slopes (Otto and Sass, 2006; Scapozza et al., 2011), this affirms the classification of FUR02 as a pebbly rock glacier. When the lobe height is transferred to the R02 radargram, however, it coincides with a pattern of surface-parallel reflectors above the undulating reflectors. This conflicting result is attributed to an unsatisfactory time-depth conversion, as this conversion is based on a constant value from literature only. The resulting uncertainty gives preference to the interpretation based on the result of the ERI survey.

Similar to the observations in the E02 model, extraordinary high resistivity values in the uppermost layer of the E03 model are attributed to accumulations of blocky materials at the risers of FUR02. In contrast to the E02 model, in which the horizontal slices show an enhancing contrast in resistivity between positions at the risers and at the tread, resistivity values of the E03 model generally increase with depth. However, it must be noted that resistivity contrasts may be unreasonably enhanced due to the applied L1-norm inversion or a prolonged inversion process (Hauck and Vonder Mühll, 2003). Despite this drawback, the L1-norm inversion was chosen to avoid problems in detecting the lower limit of permafrost bodies (Lewkowicz et al., 2011). Considering the assumed occurrence of permafrost, the small accumulation of relatively high resistivity values in the frontal part of the tread (E03: J) is interpreted to represent a small ice lens.

Chapter 8

Results and Interpretation OVR

Geophysical investigations at OVR comprise multi-dimensional approaches of ERI (8.1) and GPR (8.2). Temperature data was recorded at the ground surface, and matric potential measurements were performed at two positions (8.3). For ground truth information, the topography of the frost table was sampled within the survey areas of E01 and R01 by frost-probing. The last part of this chapter comprises an interpretation of the results, focusing the internal structure of the investigated palsas and the subsurface conditions of the study site (8.4).

8.1 Electrical Resistivity Imaging

The presented results of geoelectrical surveying are divided into two parts: those from the southern and central part of the study site are presented in the first part of this section and those from the northern part of the study site in the second part.

Southern and Central Palsa Area

The southern and central part of the study site was investigated by two r-3-D surveys (E01, E02) and four 2-D ERI surveys (E03–E06). Horizontal model slices of the E01 r-3-D ERI model (fig. 8.1a) show a patchy resistivity distribution in the subsurface. Values of up to 1.9 kΩm appear in the center of the model (A), at a position that corresponds to the center of OVR01. This structure of relatively high resistivity values is covered by a layer with 0.6 m thickness and extremely low resistivity values (<0.5 kΩm). Maximum resistivity values appear between depths of 1.2 m and 2 m. With increasing depth, resistivity values decrease to 1.5 kΩm at a depth of 5 m. Two additional structures of relatively high resistivity values are present in the northeastern (B) and the northwestern (C) corner of the model.

At spot A, resistivity values between 1.4 kΩm and 1.8 kΩm appear directly below the surface. Below a depth of 2 m, resistivity values decrease with increasing depth, but are still above 0.6 kΩm throughout the rest of the model. Structures B and C reach through the entire depth of the model. While resistivity values of structure B constantly decrease with increasing depth, resistivity values of structure C increase from 0.7 kΩm directly below the surface to 1 kΩm at a depth of 2 m. Below a depth of 2 m, resistivity values of structure C slightly decrease with increasing depth and the structure attaches to structure A in the central part of the model. In contrast to the central part of the model, where the outline of structure A corresponds to the outline of the palsa, the horizontal extents of the structures B and C do not correspond to any visible surface features. The positions of both structures of relatively high resistivity correspond to the grassy surroundings of the palsa, where only a few, relatively small mounds occur.

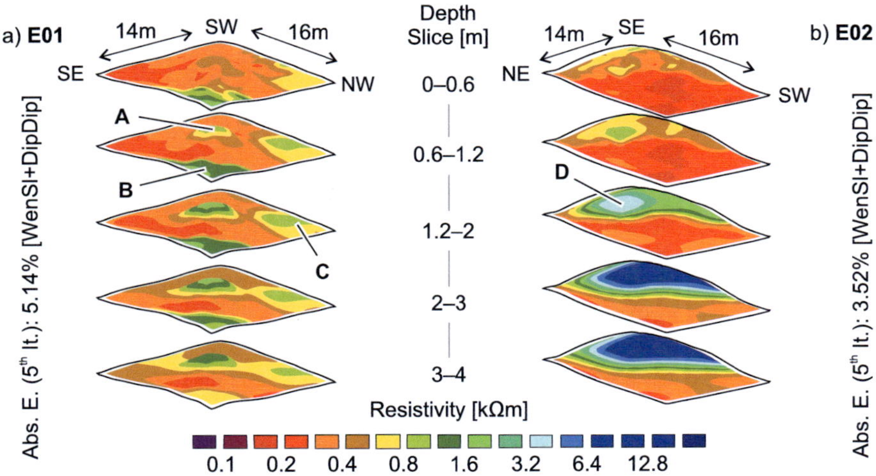

Figure 8.1: ERI Models OVR: a) E01, b) E02. Selected Slices. Labels Refer to Structures Mentioned in the Text.

Relatively low resistivity values dominate the horizontal resistivity distribution in the upper two slices of the E02 model (fig. 8.1b), and only a small gradual increase in resistivity towards the southeastern corner is visible. Below a depth of 1.2 m, the contrast in resistivity between the southeastern corner, where cells reach up to 5 kΩm, and the rest of the model, where resistivity values are constantly below 0.3 kΩm, increases strongly. At a depth of 2 m, a triangular structure of relatively high resistivity values (up to 11 kΩm) appears (D) at a depth of 3 m. With further increasing depth, resistivity values change only slightly and at the bottom of the model, values are still above 9 kΩm (not show). The shape of structure D and its position correspond to the extent and position of palsa OVR02.

In the adjacent or partly overlapping models E03 (fig. 8.2a), E04 (fig. 8.2b) and E05 (fig. 8.2c), relatively high resistivity values cluster at four positions (E–H). However, the extent of the four structures and the modeled resistivity values differ between the models (see tab. 8.1). These variations are attributed to geometrical variations between the survey positions, differences in the electrode spacing and temporal alterations between the survey dates (see tab. A.4, appendix). The structures are embedded in a matrix of values between 0.1 kΩm and 0.5 kΩm. While structures E, G and H show discrete lens-shaped outlines, structure F is displayed as two separated structures in the E03 model. The same model indicates further that the structures E and F and the structures at G and H are connected by bands of resistivity values between 0.5 kΩm and 0.6 kΩm at a depth of around 4 m. Except for the appearance of structure F in the E03 model, all structures of relatively high resistivity values are covered by a layer of relatively low resistivity values (<0.4 kΩm).

In addition to the variations in extent and resistivity, the height of the uplifted areas above the four structures differs: At spot E, the extent of the resistivity structure exceeds the uplifted area (OVR02) in both horizontal dimensions. The extent of the relatively small structure F is almost equal to the extent of the uplifted area (OVR01). The surface above the more extensive structure G is only barely uplifted at isolated positions. At spot H, the extent of the uplifted area is nearly equal to the extent of the corresponding structure of relatively high resistivity.

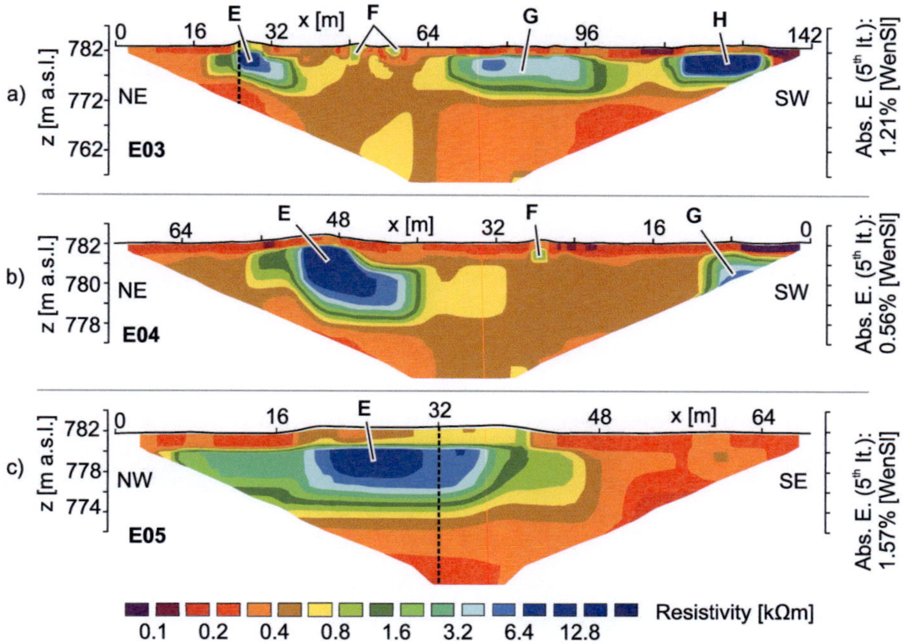

Figure 8.2: ERI Models OVR: a) E03, b) E04, c) E05. Labels Refer to Structures Mentioned in the Text.

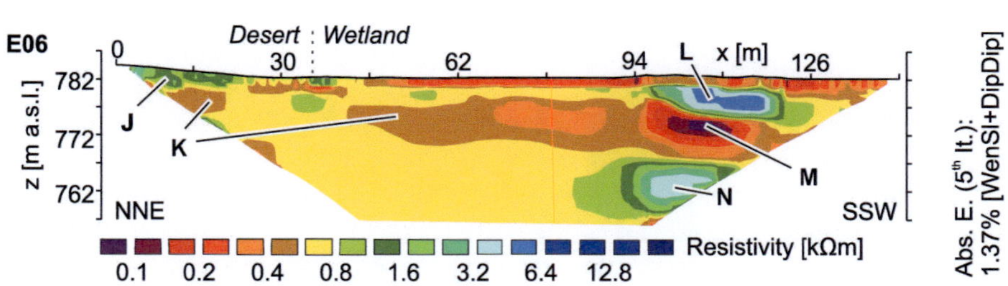

Figure 8.3: ERI Model OVR: E06. Labels Refer to Structures Mentioned in the Text.

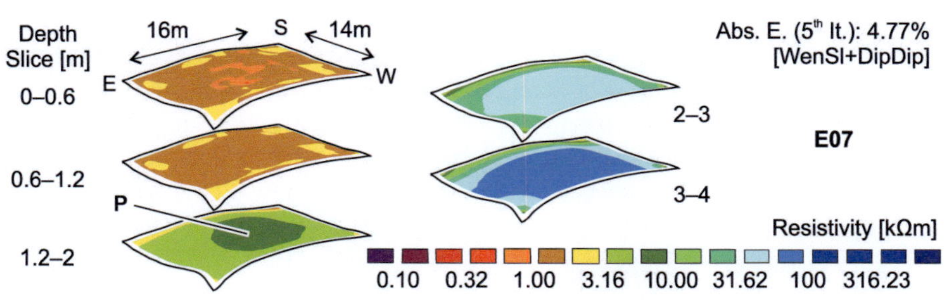

Figure 8.4: ERI Model OVR: E07. Selected Slices. Labels Refer to Structures Mentioned in the Text.

Table 8.1: Differences in Thickness and Resistivity Between Identical Structures in Different ERI Models (OVR).

	E (E03)	F (E03)	G (E03)	H (E03)	E (E04)	F (E04)	G (E04)	E (E05)
ρ_{max} [kΩm]	7.5	1	5	16.5	10.5	5.5	1	10
t High Res.	8.6	4.3	8.6	6	5.6	–	0.7	4.3
t Low Res.	–	–	1.4	1.4	1	2.2	1.5	2

t High Res. = Thickness of Structure of High Resistivity
t Low Res. = Thickness of Covering Layer of Low Resistivity

The E06 model (fig. 8.3) provides information on the subsurface conditions at the transition between the wetland area and the surrounding desert. The part of the model that corresponds to the surrounding desert shows an upper layer with around 3 m thickness and relatively high resistivity values between 0.9 kΩm and 2.4 kΩm (J). Below this layer, resistivity values decrease to around 0.5 kΩm (K). While the uppermost layer J is visible only in the part of the model that corresponds to the surrounding desert, layer K can be traced throughout the complete length of the model, except for an interruption of around 19 m length around the transition between the two parts of the model.

Relatively high resistivity values of up to 5.8 kΩm cluster in a structure of up to 6 m thickness (L). The extent of this structure corresponds to an area where two adjacent palsas form an uplifted area of around 0.6 m height. Structure L is covered by a layer with 1.8 m thickness and resistivity values below 0.5 kΩm. Below structure L, resistivity values decrease to a minimum of 0.1 kΩm (M) before they increase again to about 5 kΩm (N).

Northern Palsa Area

Palsas of the northern part of the study site were investigated by three 2-D ERI surveys and one r-3-D ERI survey. The r-3-D ERI model E07 (fig. 8.4) corresponds to the subsurface of OVR03 and shows a rather uniform resistivity distribution pattern, compared to the r-3-D ERI models from the central part of the study site (E01, E05). The upper two layers of the model show resistivity values between 0.8 kΩm and 2.1 kΩm. In the following depth slice, resistivity values increase sharply to around 8 kΩm at the center of the model (P). With increasing depth, resistivity values at spot P reach up to 48 kΩm (depth slice 2 m–3 m) and 89 kΩm (depth slice 3 m–4 m), respectively. Between a depth of 4 m and the bottom of the model at a depth of 5 m, resistivity values reach up to 100 kΩm (not shown).

The 2-D ERI model E08 (fig. 8.5) corresponds to a cross-section of the palsas OVR03 and OVR04 and the transitional area in between. It shows two structures of relatively high resistivity values (Q, R): the position of structure Q, which shows resistivity values of up to 98 kΩm, corresponds to the subsurface of OVR03 and the position of structure R, which shows resistivity values of up to 89 kΩm, corresponds to the subsurface of OVR04. The two structures Q and R are separated by a zone of extremely low resistivity values between 0.01 kΩm and 0.4 kΩm (S). This zone stretches over a length of 12 m and its position corresponds to the transitional area between the two palsas. However, only a small part of the model corresponds to the actually water-covered area. Where the model corresponds to the subsurface of the two palsas, resistivity values sharply increase below depths of 1.3 m and 1.9 m. How-

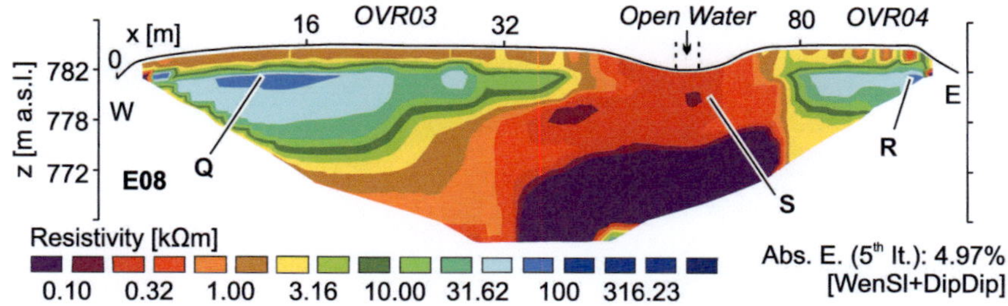

Figure 8.5: ERI Model OVR: E08. Labels Refer to Structures Mentioned in the Text.

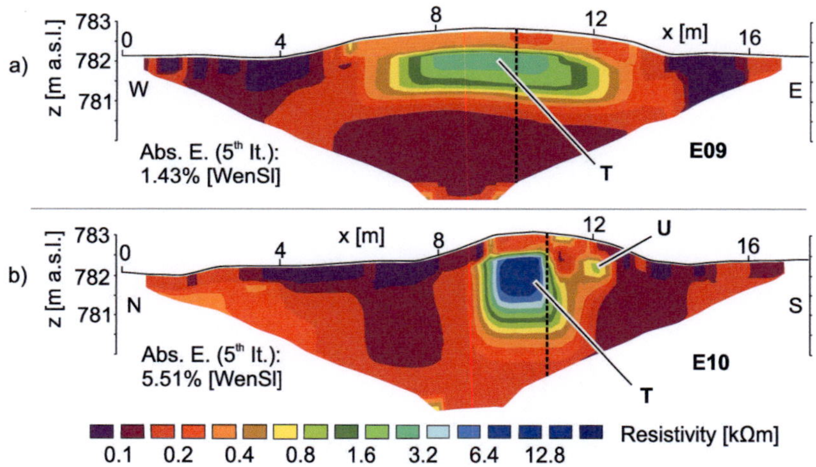

Figure 8.6: ERI Models OVR: a) E09, b) E10. Labels Refer to Structures Mentioned in the Text.

ever, resistivity values at both sides of this boundary are above 1 kΩm. A lower boundary of the structures Q and R is detected at depths between 10 m (Q) and 6 m (R), respectively.

The intersecting 2-D ERI models E09 (fig. 8.6a) and E10 (fig. 8.6b) correspond to cross-sections along the longitudinal (E09) and the transverse (E10) axis of OVR05, respectively. However, both models show distinct differences in the observed resistivity distribution. The E09 model shows a lens-shaped structure of 0.9 m thickness and relatively high resistivity values between 0.6 kΩm and 3 kΩm (T). The structure is covered by a shallow layer of resistivity values below 0.4 kΩm. The E10 model shows a similar thickness of structure T at the intersecting position with E09, but markedly higher resistivity values of up to nearly 9 kΩm. Thickness and resistivity values of the shallow cover layer, however, are similar in both models. The horizontal extent of structure T is 5.5 m in longitudinal direction (E09) and 2.5 m in transverse direction (E10). Only the transverse E10 model but not the longitudinal E09 model reveals the appearance of an additional patch of relatively high resistivity values (U) that is separated from structure T and shows resistivity values of up to 0.8 kΩm. Between the separated patch and the main structure, the thickness of the upper layer of relatively low resistivity values increases from 0.4 m to 0.9 m. This part of the model corresponds to the location of a crack on the palsa surface (see fig. 2.6d).

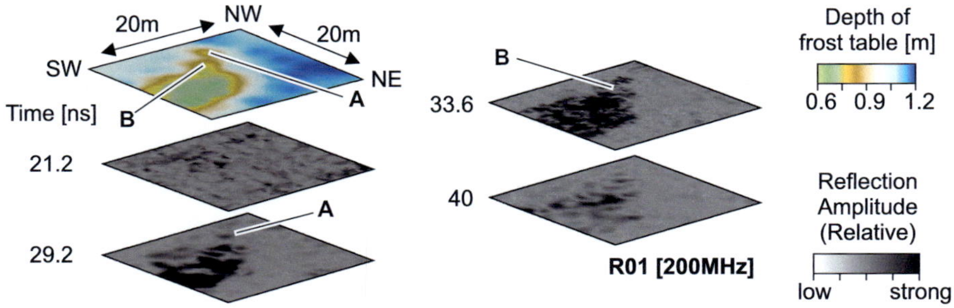

Figure 8.7: Radargram OVR: R01 and Frost Table Topography in Corresponding Area. Selected Depths Slices. Labels Refer to Structures Mentioned in the Text.

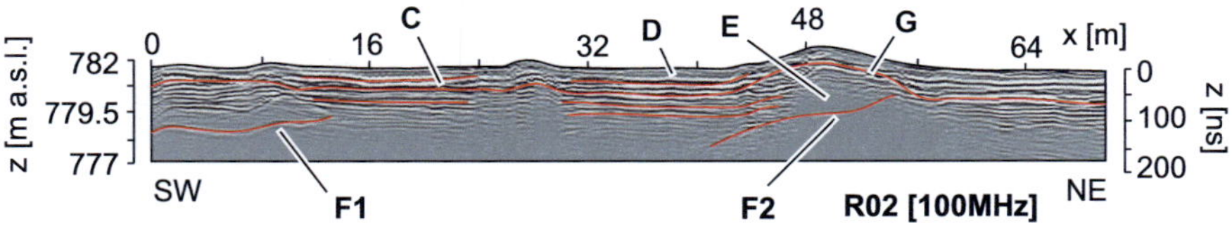

Figure 8.8: Radargram OVR: R02. Labels Refer to Structures Mentioned in the Text. Selected Reflectors are Accentuated in Red.

8.2 Ground-Penetrating Radar

Besides five 2-D GPR surveys, one 3-D GPR survey (R01) was performed to assess the vertical EM velocity layering. Therefore, a rather innovative approach was applied, making use of the appearance of a strong reflector in the horizontal slices of the 3-D GPR model R01. This reflector resembles the shape of the frost table below OVR02, as derived from manual frost-probing (fig. 8.7). When the TWT until the initial appearance of the detected reflector is adjusted to the measured depth of the frost table at positions of the investigated area, an EM velocity of 0.05 m ns^{-1} is derived: this value matches a frost table depth of 0.7 m to a TWT of 28 ns (A) and a frost table depth of 0.9 m to a TWT of 36 ns (B). Although this value actually represents subsurface conditions above the frost table only, it was used as a constant value for time-depth conversion of the 2-D GPR survey data.

Radargram R02 (fig. 8.8) corresponds to a cross-section of the central palsa area and shows different reflection patterns. A pattern of surface-parallel reflectors (C) is covered by a shallow layer with weak reflection amplitude (D). Pattern C reaches into a depth of around 3 m at positions that correspond to the areas between the palsas but is reduced to a depth of around 1 m at positions that correspond to the uplifted areas. Where the depth of pattern C is reduced, blank patches appear (E1–E3). However, two of these blank patches are vertically divided by single reflectors (F1, F2). Among the surface-parallel reflectors of pattern C, one prominent, wavy reflector is continuously traceable through the entire model at depths between 0.7 m and 1.3 m (G).

The first part of the R03 radargram (fig. 8.9) correspond to the subsurface of the surrounding desert and shows a pattern of strong, slightly descending reflectors directly below

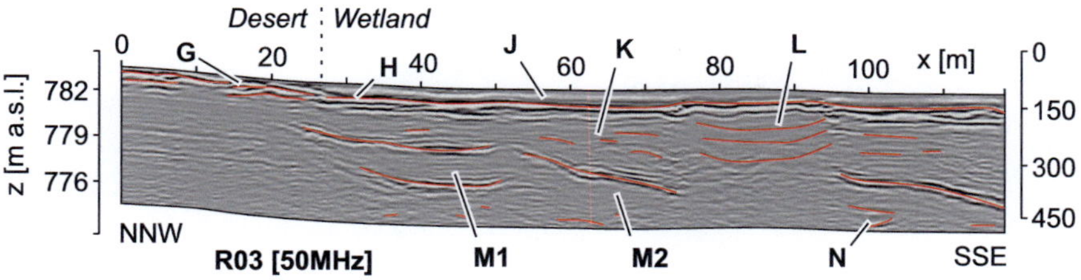

Figure 8.9: Radargram OVR: R03. Labels Refer to Structures Mentioned in the Text. Selected Reflectors are Accentuated in Red.

the ground surface (G). Only in the following part, one prominent reflector (H) forms a distinct boundary between an upper layer of relatively weak reflection amplitude (J), and a chaotic pattern of short reflectors (K). This boundary is located at a nearly constant depth level of around 1 m and rises only between $x = 75$ m and $x = 95$ m. In this area, which corresponds to the subsurface of an unnamed palsa, pattern K is replaced by blank spaces between concave reflectors (L). At the bottom of the radargram, prominent descending or concave reflectors appear, which partially enclose blank areas (M1, M2). At the bottom of the radargram, a chaotic reflection pattern is visible (N).

The R04 radargram (fig. 8.10a) corresponds to a longitudinal cross-section of OVR05. It shows a pattern of surface-parallel reflectors and an increase in reflection amplitude from the western to the eastern part of the radargram. Two prominent reflectors appear in the central part of the radargram, which corresponds to the extent of the palsa: one surface-parallel reflector is located at a depth of around 0.5 m (Q) and the other at a depth of only a few centimeters below the surface (R). The two prominent reflectors Q and R also appear in the perpendicular R05 radargram (fig. 8.10b) where, likely due to a higher reflection amplitude compared to the R04 radargram, they can be traced throughout the complete radargram. Although the survey line of R05 crossed the surface crack ($x = 12$ m), the R05 radargram appears unaffected from this feature.

The R06 radargram (fig. 8.10c) corresponds to a cross-section of OVR04. At a depth of 0.7 m, the radargram shows a prominent reflector (S) that follows the course of the surface topography. After around $x = 33$ m, however, the descend of this reflector S is stronger than the descend of the surface and therefore, the reflector descends to a depth of 0.9 m at the end of the radargram. The appearance of a second surface-parallel, but rather weak reflector at a depth of 0.2 m (T), is limited to the first part of the radargram, which corresponds to the subsurface of the palsa (T).

8.3 Temperature and Water Potential Measurements

At the sensor position of the surround desert (T01), daily mean temperature values between 19.5 °C (Jul 25, 2017) and −16.5 °C (Jan 20, 2016) were recorded. Daily mean values fluctuated strongly throughout the complete investigated time span and the curve of recorded GST values lacks durable periods of zero-curtain conditions (fig. 8.11a). A similar temperature curve, but with slightly damped values and a short period of zero-curtain conditions,

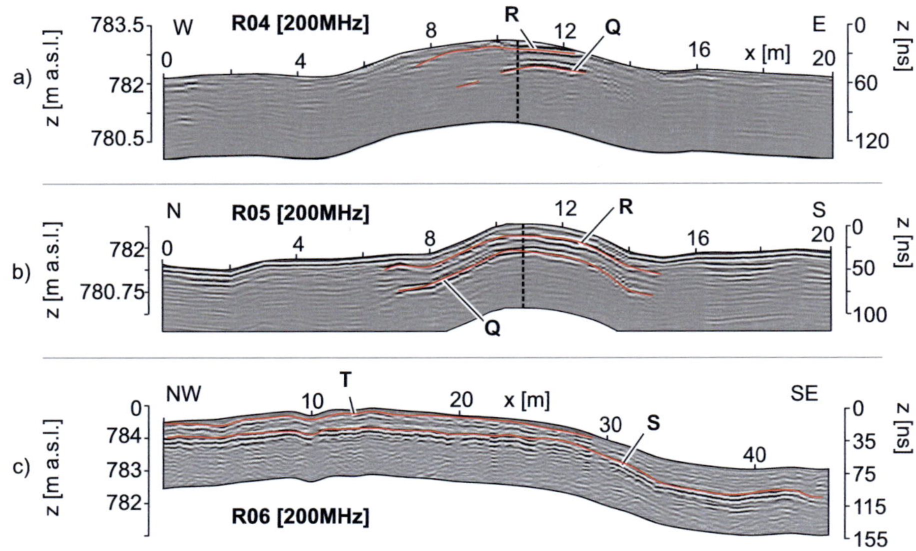

Figure 8.10: Radargrams OVR: a) R04, b) R05, c) R06. Labels Refer to Structures Mentioned in the Text. Selected Reflectors are Accentuated in Red.

is derived from the GST records at T02 (fig. 8.11b). At this sensor position on the surface of OVR04, higher minimum values (−14.3 °C) and lower maximum values (14.0 °C) are recorded. At both positions (T01, T02), extreme values were recorded at the same date.

The GST plot of sensor T03 (fig. 8.11c), located in the wetland area between two palsas, resembles the curves of T01 and T02 only during summer. Therefore, only the maximum daily mean temperature value of 12.5 °C was recorded at the same date as the maxima of the other sensors were recorded ($Tmin = -2.74$ °C, Dec 14, 2015). As indicated by the relatively high minimum temperature value, extremely cold winter temperatures are absent. Between March and May, the GST plot of T03 shows a relatively long period of zero-curtain conditions. Between the spring of 2015 and the spring of 2016, the number of days with zero-curtain conditions decreased at the sensor locations of T02 and T03, while the number slightly increased at T01. Interannual differences in MAGST values between the sensor locations are in a similar range, with an average of −1.22 K (tab. 8.2).

The curves of subsurface temperature values and matric potential values at T02 (fig. 8.11e) can be divided into four phases: Between mid-October 2015 and mid-December 2015, subsurface temperature values were nearly constantly around 0 °C at a depth of 0.55 m (Phase A). Phase A ended in mid-December 2015 with a nearly simultaneous increase of pF-values and decrease of temperature values. It was followed by a phase of high pF-values and sub-zero temperatures (Phase B). Short-term temperature fluctuations in this period coincided with fluctuations of pF-values. Phase B lasted until early April 2016 and was followed by a transitional phase (Phase C). Phase C was characterized by gradually increasing temperatures that approached a value of 0 °C, and by gradually decreasing pF-values. The next phase (Phase D) started in early July 2016, when daily mean temperatures started to be constantly positive and pF-values reached a minimum level, on which they remained for the next six months.

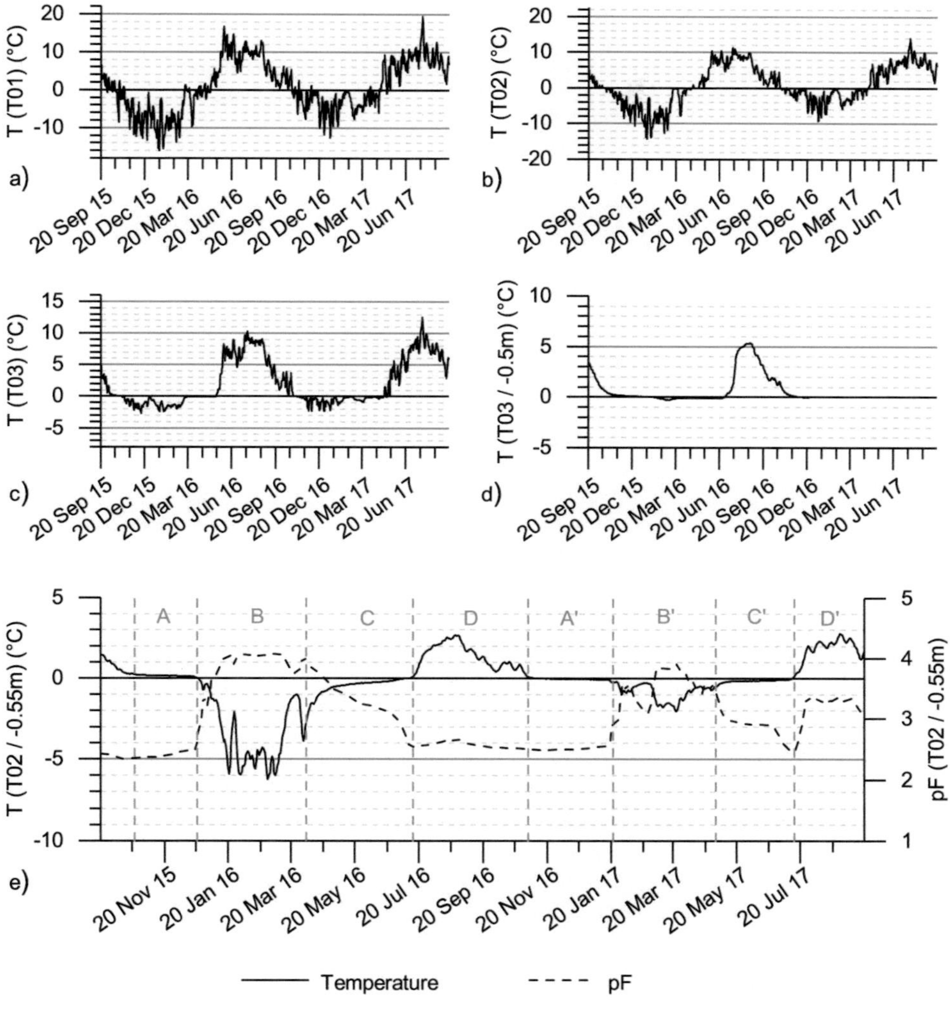

Figure 8.11: Recorded Data OVR: a)–d) GST, e) pF-Values.

After phase D, which started in early November 2016, this cycle was repeated (A'–D'). Variations between the two cycles mainly concern the magnitude of recorded temperature and matric potential values. One more distinct difference between the two years of investigation concerns the beginning of the phases D and D', respectively: in the second cycle, pF-values increased exceptionally strong after mid-July 2017.

The subsurface sensor at T03 failed on Dec 24, 2016. Until this day, the logger had recorded a constant pF-value of 0 throughout the complete investigated time span (not shown). The corresponding plot of daily mean subsurface temperature values (fig. 8.11d) shows only low fluctuations. Positive temperature values were recorded throughout nearly the complete recorded time span and thus, a maximum frost depth of 0.5 m is assumed.

Table 8.2: Summary of Recorded GST Data: OVR.

Label	MAGST [°C] 2015/2016*	MAGST [°C] 2016/2017*	SC [d] 2015/2016*	SC [d] 2016/2017*	Topographic Position
T01 Surface	0.3	1.56	0	10	Surrounding Desert
T02 Surface	0.25	1.52	54	41	Surface of OVR04
T03 Surface	1.09	2.21	186	160	Wetland Area
T02 (–0.55 m)	–0.67	0.16	–	–	Subsurface of OVR04
T03 (–0.5 m)	0.94	No data	–	–	Subsurface of Wetland Area

MAGST = Mean Annual Ground Surface Temperature // SC = Snow Cover
*Investigated Period: 20 Sep–19 Sep

8.4 Interpretation

Despite relatively high MAGST values, the curves of the recorded GST values agree with the assumption of permafrost-favorable conditions at OVR, as stated by Saemundsson et al. (2012) and Hirakawa (1986). Strong diurnal temperature fluctuations in the recorded data from T02 indicate that the surfaces of the palsas are only slightly covered with snow during winter. The existence of an insulating snow cover is assumed for the areas between the palsas from the records of T03. Snow-free conditions at the surface of OVR04 are presumably responsible for the shallow active layer thickness, which was detected only by 2-D GPR (R06: S) and ranges between 0.7 m and 0.9 m. The appearance of resistivity values of up to nearly 100 kΩm at OVR03 (E08: Q) and 90 kΩm at OVR04 (E08: R) indicates the occurrence of massive segregation ice. However, the magnitude of resistivity may be enhanced by the enormous palsa height and a subsequent gravitational loss of liquid water in the uplifted layers. The upper reflector in the R06 radargram (R06: T) is interpreted as a boundary between different types of subsurface materials, like the boundary between layers of sandy loam and loam as reported by Saemundsson et al. (2012).

The short time lag between changes in subsurface temperature and matric potential at T02 shows a close and direct coupling between the two parameters. The observed cycle highlights the ability of liquid water to buffer temperature variations in the active layer. Decreasing pF-values in spring show that subsurface materials can hold a relatively high amount of liquid water, although negative temperature values indicate a frozen state. The strong increase in pF-values in July 2017 is attributed to a period of extraordinary high evaporation, which led to desiccation in this part of the active layer, as all GST sensors show their maximum temperature values around this time of the year. Positive subsurface temperature values and the absence of any pF-value variations at T03 show that the subsurface at this position is permanently in an unfrozen, water-saturated state. However, it must be noted that the lack of any fluctuations from the displayed pF-value of 0 is suspicious and may indicate malfunction even in advance of the complete sensor failure in December 2016.

GST records from the surrounding desert (T01) indicate snow-free conditions outside the wetland area, likely due to snow redistribution by wind. This concept further explains the strong variations of MAGST values between the different sensor positions and the inconsistent development of the zero-curtain period. Despite the relatively cold temperatures, a

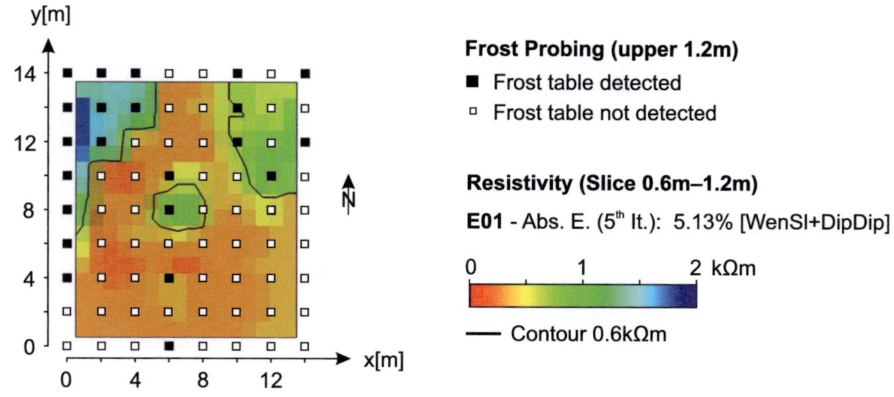

Figure 8.12: Overlay Analysis: Manually Detected Frost Table Depth and ERI Model E01 (OVR).

possible occurrence of permafrost is unlikely outside the wetland area. This is because frost susceptibility of the rather coarse-grained materials of the surrounding desert is lower than frost susceptibility of the fine-grained materials in the wetland area (Chamberlain, 1981). However, the nearly continuous layer of relatively low resistivity values in the corresponding ERI model (E06: K) indicates that coarse materials only form a covering layer.

A comparison between the horizontal resistivity distribution of the E01 model and the manually sampled frost table topography provides information on the relationship between resistivity and permafrost: highest conformity between the two data sets is achieved by a threshold value of 0.6 kΩm to distinguish between frozen and unfrozen conditions, as this value can reproduce frost-probing results at 37 of the 42 sample locations (fig. 8.12). Although this approach provides only a rough approximation due to differences in the vertical resolution of ERI and frost probing, it can still be used for interpretation purposes.

Structures of relatively high resistivity values in the central part of the study site, which partially correspond to areas without any GPR reflections, indicate the occurrence of ice layers. However, 2-D ERI models from this area reveal discrepancies between the extent of the uplifted areas and the extent of the frozen cores. A comparison between the 2-D ERI models and the R02 radargram reveals that the presumably frozen structures differ by appearance (E03–E05: E, G) or absence (E03–E04: F) of horizontal reflectors that divide these structures (R02: F) into compound features. The appearance of surface-parallel reflectors between the ice layers (R02: C) indicates an aeolian deposition of the subsurface materials (Van Overmeeren, 1998). This aligns with the high rates of this type of deposition in central Iceland (Arnalds, 2010). The continuous reflector in the R02 radargram (R02: G) is assumed to represent a continuous frost table, as its course corresponds with the course of the manually detected frost table. Extensively frozen subsurface conditions are indicated by the connecting bands of relatively high resistivity values between the structures E and F and between the structures G and H in the E03 model, when the assumed threshold value of 0.6 kΩm is assumed. The absence of the connection between the structures E and F in the E04 model indicates that this connection disintegrated between 2015 and 2017. However, it must be noted that the models E03 and E04 have a different spatial resolution and that surveying was not performed at exactly the same position.

Generally higher resistivity values of the E06 model, compared to the other resistivity models, indicate different subsurface conditions at the margin of the wetland area, compared to the central part. Results of both geophysical surveys (E06, R03) clearly reflect the transition between the wetland area and the surrounding desert (E06: J, R03: G). The boundary (R03: H) that separates an upper part with weak reflection amplitude and relatively low resistivity values from a lower part with a chaotic reflection pattern and relatively high resistivity values, is interpreted as the frost table. It is located at depths between 1 m and 3 m. Blank spaces in the radargram (R03: M1, M2), as well as isolated structures of relatively high resistivity values (E06: N) and a nearly continuous layer of relatively low resistivity values (E06: K) in the ERI model indicate a rather complex subsurface layering. It likely includes alternating layers of frozen and unfrozen conditions, as well as the occurrence of massive ice at the bottom of the water-saturated layer. The high magnitude of the vertical resistivity alternation in the E06 model (E06: L–M–N) may be exaggerated by a prolonged inversion process, but its basic appearance already in the 2nd inversion step shows that this feature is backed by the data and not an artifact (not shown).

The two 2-D ERI models from OVR05 are characterized by strong differences in the resistivity distribution between the longitudinal (E09) and the transverse (E10) model. Therefore, the two models must be interpreted carefully. The assumed thickness of the frozen core (E09, E10: T) varies between 1.5 m (E09) and 2 m (E10), but it must be noted that 2-D ERI may generally underestimate the thickness of a permafrost layer (Lewkowicz et al., 2011). Connecting bands of relatively high resistivity values to other palsas or to the base of the water-saturated layer are unlikely, as a distinct lower boundary of the frozen core is visible in both models. The patterns of surface-parallel reflectors, which are visible in the radargrams R04 and R05, resemble the pattern of the R02 radargram, which corresponds to a cross-section of the central part of the wetland area (R02: C). While reflectors in the upper subsurface (R04, R05: R) are attributed to aeolian deposition (Van Overmeeren, 1998), the lowermost, prominent reflector is assumed to represent the frost table (R04, R05: Q). This is affirmed by an increase in resistivity at a similar depth in the corresponding 2-D ERI models.

Chapter 9
Results and Interpretation HPA

Geophysical investigations at HPA comprise multi-dimensional approaches of ERI (9.1) and 2-D GPR (9.2). Additionally, temperature data was recorded at the ground surface (9.3). The last section of this chapter presents an interpretation of the results, focusing on the internal structure of the investigated patterned ground phenomena (9.4).

9.1 Electrical Resistivity Imaging

The q-3-D ERI model E01 (fig. 9.1a) shows the resistivity distribution in the subsurface of HPA01. The upper two model slices show a patchy resistivity distribution, which reflects the grain-size distribution of the surface: Relatively low resistivity values below 2 kΩm correspond to slightly vegetated parts of the survey area, where the material composition comprises a great variety of different grain sizes (A) and relatively high resistivity values between 2 kΩm and 8 kΩm correspond to areas where fine-grained materials dominate (B). This observation shows the sensitivity of the q-3-D ERI approach and is enabled by the high density of data points (see tab. A.1, appendix), which results from using a survey setup with both an electrode and line spacing of 2 m over the complete survey area.

Below a depth of 1.5 m, resistivity values increase throughout the complete horizontal extent of the E01 model. Highest values accumulate at two spots (C, D): in the central part of the model (C), values reach up to 28 kΩm between depths of 1.5 m and 2.5 m, while values at the eastern margin of the model (D) reach up to 58 kΩm between depths of 2.5 m and 3.5 m. The model shows decreasing resistivity values with increasing depth and the outlines of the two structures with high resistivity values C and D vanish below a depth of 5 m. The lowermost model slice, which represents a depth level between 6.5 m and 8 m, shows a relatively uniform resistivity distribution with values between 2 kΩm and 12 kΩm.

The intersecting 2-D ERI model E02 (fig. 9.1b) shows a similar resistivity layering. A layer of relatively low resistivity values between 1.5 kΩm and 8 kΩm appears in the upper 1.5 m of the subsurface (E), followed by a layer of resistivity values between 11 kΩm and 47 kΩm (F). The E02 model shows the occurence of two additional clusters of relatively high resistivity values in this layer, which appear outside the overlapping part (G, H). With further increasing depth, the resistivity values in the southeastern part of the model decrease (J).

The r-3-D ERI model E03 (fig. 9.2) corresponds to the subsurface of HPA02 and shows only small variations in the horizontal resistivity distribution. A layer with relatively low resistivity values, mainly between 1 kΩm and 2 kΩm, reaches down to a depth of 1.2 m (K). Below this layer, resistivity values increase sharply from 2 kΩm to 7 kΩm. Only in the southern part of the model, relatively low resistivity values, below 4 kΩm, are visible throughout the entire model depth of 5 m (L).

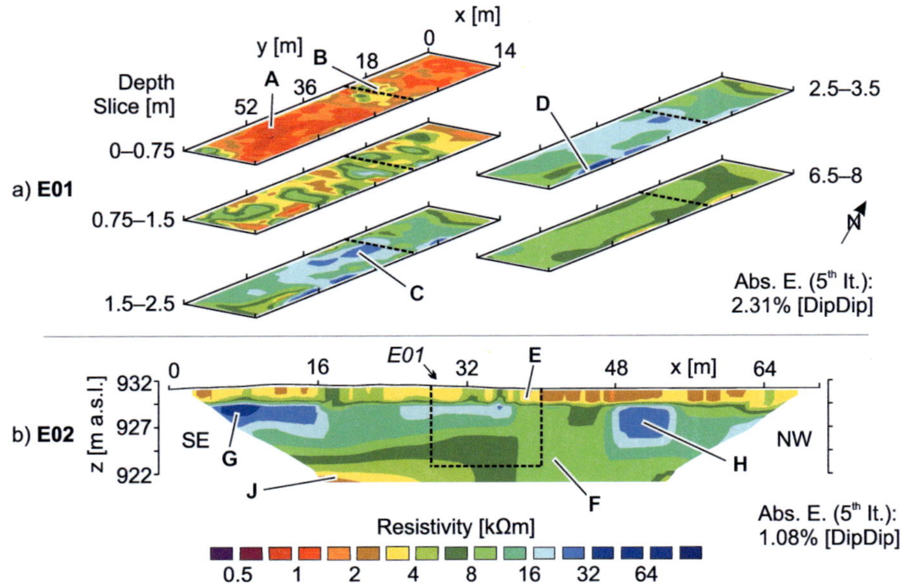

Figure 9.1: ERI Models HPA: a) E01, b) E02. Selected Slices. Labels Refer to Structures Mentioned in the Text. Dashed Line/Box Mark Intersection between the Models.

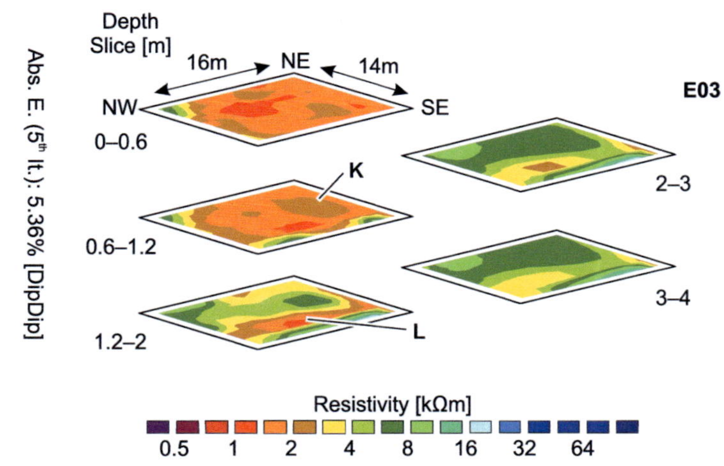

Figure 9.2: ERI Model HPA: E03. Selected Slices. Labels Refer to Structures Mentioned in the Text.

9.2 Ground-Penetrating Radar

For time-depth conversion of the data from the 2-D GPR surveys at HPA01 (R01–R03) and at HPA02 (R04) a constant EM velocity value of 0.08 m ns^{-1} was used, based on the detection of a few diffraction hyperbolas in the radargrams. The R01 radargram (fig. 9.3a) corresponds to a parallel cross-section of the survey line of E02. It shows a structure of four units: Down to depths of around 1 m, the radargram lacks any prominent reflectors (A). Below this layer, the radargram shows a pattern of curved reflectors, that reach down to a depth of around 5 m (B). Below this depth, the reflection amplitude decreases significantly (C). Reflection amplitude increases again only towards the bottom of the radargram, where a pattern of curved reflectors is visible (D). The upper boundary of pattern D shows a slight descend in survey direction.

The upper layer without prominent reflectors A is repeated in the perpendicular R02 radargram (fig. 9.3b). However, in contrast to the R01 radargram, the lower boundary of this layer A is formed by one prominent, continuous reflector at a depth of around 3 m (E). Below this reflector, reflection amplitude is low (C). A high reflection amplitude is restored only in the lowermost part of the radargram (D). The upper boundary of pattern D shows again a descending course in survey direction. Except for the differences in the upper part (B/E), the radargrams R01 and R02 show a congruent layering at the intersection. The R03 radargram (fig. 9.3c) shows again the structure of the parallel R01 radargram (A–D). Only the course of the boundary between the two lowermost reflection patters differs: while it has an inclined orientation in the radargrams R01 and R02, it is located at a constant depth of around 10 m in the R03 radargram.

A different subsurface layering is observed in the R04 radargram (fig. 9.4). The radargram shows multiple prominent reflectors, embedded in a chaotic pattern of rather weak reflectors. One prominent reflector (F) appears in the first part of the radargram and declines from the surface to a depth of around 1.5 m at $x = 10.5$ m. It is replaced by an 11 m long, horizontally striking reflector at a depth of about 0.6 m (G) and by a short, inclining reflector at a depth of about 2 m (H). The northeastern part of the radargram shows a pattern with multiple inclining reflectors (J).

9.3 Temperature Measurements

Recorded temperature data shows a MAGST of 1.76 °C for the complete year-round period from October 01, 2015 to September 30, 2016. The temperature curve (fig. 9.5) shows highly fluctuating daily mean values after the start of the investigation period. An overall minimum value of daily mean temperatures of −3.55 °C was recorded on November 20, 2015. Diurnal temperature fluctuations were below 0.4 K on 193 days in the winter of 2015/2016 and indicate a durable, insulating snow cover. The occurrence of such a snow cover is affirmed by a relatively long zero-curtain period of 74 days, that lasted until March 2016.

After the zero-curtain period in spring 2016, temperature values increased by up to nearly 6 K within only two days, and a temperature of 11 °C was reached in Mid-July 2016. In the subsequent weeks, temperature values decreased and reached a level of 0 °C at the end of October 2016. In the following winter of 2016/2017, daily mean temperature values fluctuated

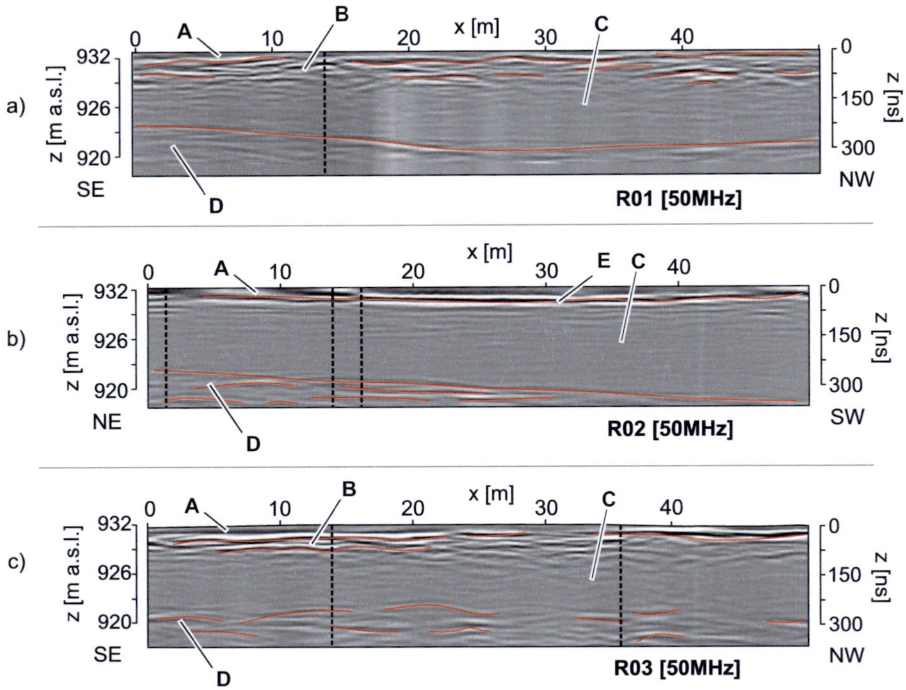

Figure 9.3: Radargrams HPA: a) R01, b) R02, c) R03. Selected Reflectors are Accentuated in Red.

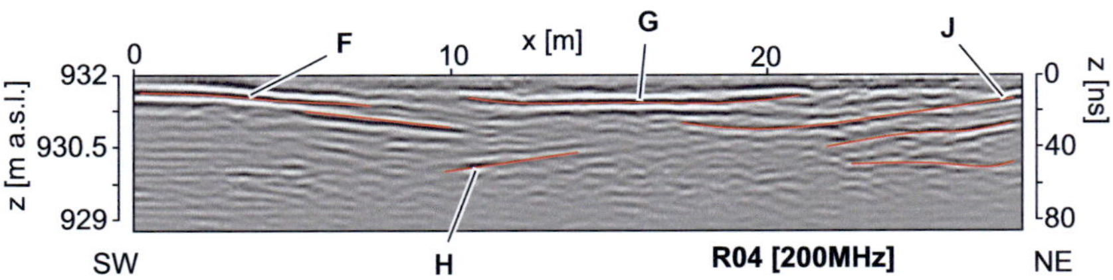

Figure 9.4: Radargram HPA: R04. Selected Reflectors are Accentuated in Red.

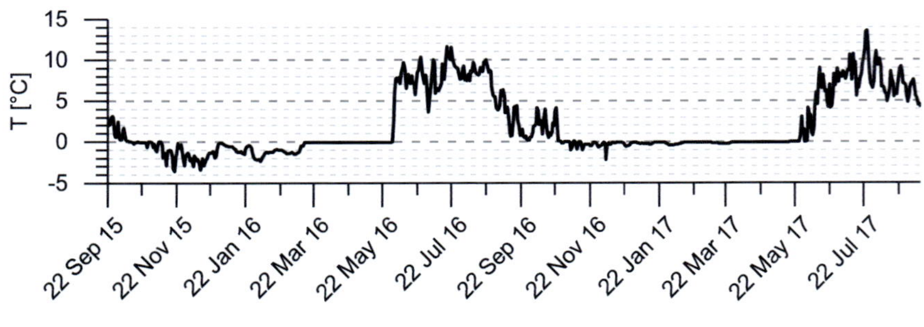

Figure 9.5: Recorded Ground Surface Temperature Data HPA.

only slightly around 0 °C. Low diurnal fluctuations indicate the existence of an insulating snow cover on 209 days in spring 2017. Like in the previous summer of 2016, temperature values started to increase from the end of May. An overall maximum value of daily mean temperature of 13.6 °C was reached on July 25, 2017.

9.4 Interpretation

Although only one complete year-round record of GST data is available, a mean temperature value of 2.13 °C from October 01, 2016 to September 10, 2017 indicates that the MAGST of 2016/2017 will be significantly higher than the MAGST of 2015/2016. This increase is attributed to a more durable snow cover in the winter of 2016/2017 which prevented the penetration of cold winter temperatures into the ground. The GST plot from HPA (fig. 9.5) indicates that the ground thermal regime is characterized by an enhanced ground cooling in early winter (Ishikawa, 2003).

Although relatively high temperature values after the onset of an insulating snow cover indicate permafrost-unfavorable conditions, the layer of relatively high resistivity values in the models from HPA01 (E01: C; D; E02: F) is interpreted as permafrost layer. This assumption is based on the observed range of resistivity values, which is not characteristic for unfrozen unconsolidated materials, and could be explained else wise only by (i) air-filled voids in the subsurface or (ii) solid rock (see tab. 3.1). Both options are conceivable, particularly as lava is exposed near the survey lines, a material which can be both massive or porous (see fig. 2.7). Although the first option can be neglected regarding the reflection patterns of the radargrams R01-R03, which lack any indication of air-filled voids, the observed low reflection amplitude in the central part of the radargrams (R01–R03: C) could be also caused by a high signal attenuation in solid rock (Sass, 2007). However, a pattern of characteristic reflectors (see e.g., Miyamoto et al., 2005), which should be traceable through the R03 radargram that corresponds to a cross-section of the lava outcrop, is absent. Assuming frozen conditions, the distinct boundary in the upper part of the R02 radargram (R02: E) reflects the thermal interface at the permafrost table (Stevens et al., 2008). The absence of prominent reflectors in the frozen layer indicates ice-saturated subsurface conditions (Moorman et al., 2003).

The r-3-D ERI model E03 from HPA02 shows a distinct boundary at depths between 1.5 m and 2 m. As the vertical resistivity gradient across this boundary resembles the gradient at the boundary between the layer E and F in the E02 model, a similar subsurface layering is observed at HPA01 and HPA02. Lower resistivity values below this boundary in the E03 model, compared to the E02 model, indicate a lower ice content at HPA02. Declining and inclining reflectors in the R04 radargram resemble the characteristic reflection patterns of glacio-fluvial or glacio-lacustrine deposits (Van Overmeeren, 1998). This type of material deposition is obvious as the study site is close to the recent margin of Hofsjökull. The course of the reflectors F, G and J of the R04 radargram delineates a surface depression of 0.3 m depth, filled with fine-grained materials. The absence of this layer in the corresponding E03 model is attributed to an insufficient vertical resolution of the applied ERI approach.

Chapter 10
Results and Interpretation SYN

This chapter presents results from the comparative analysis between resistivity models created with the software products BERT and RES3DINV, respectively. To differentiate between the reference data and the final models, the terms *reference model* and *inversion model* are used in the following. Inversion models of two reference data sets are compared, which represent a rock glacier (SYN01, see sect. 10.1) and a palsa area (SYN02, see sect. 10.2).

10.1 Rock Glacier Model

Both inversion models (fig. 10.1) show relatively low misfit errors of 6.26 % (BERT) and 4.54 % (RE3DINV), respectively. The number of cells is lower in the BERT inversion model (9073) compared to the RES3DINV inversion model (16835). Horizontal slices of both inversion models show the appearance of an elongated structure of relatively high resistivity values (A) at the depth slice of 0.5 m. This position is shallower than the position of the corresponding structure in the reference model SYN01 (fig. 3.4), which appears only below a depth of 1 m. The uniform resistivity distribution of the uppermost layer of the reference model is not resolved in both inversion models. The horizontal depth slice of 2 m shows higher resistivity values for structure A in the RES3DINV inversion model, which is closer to the values of the corresponding structure in the reference model. Below a depth of 10 m, a narrow band of relatively low resistivity values (B) appears in the horizontal slices of both inversion models. This erroneous structure appears more clearly in the RES3DINV inversion model.

A quantitative comparison between the distribution of resistivity values shows a wider range of resistivity values in the BERT inversion model. In this inversion model, single outliers reach values of up to 130 000 kΩm (fig. 10.2a). The minimum resistivity value of both inversion models of 0.05 kΩm, however, is present in the RES3DINV inversion model. The median of modeled resistivity values is higher in the RES3DINV inversion model (7.9 kΩm), compared to the BERT inversion model (4.1 kΩm). The relative histogram (fig. 10.2b) shows a similar frequency distribution of resistivity values in the inversion models from both software products. Resistivity values between 1 kΩm and 2 kΩm dominate both inversion models, but this class is more frequent in the BERT inversion model. The class of resistivity values between 20 kΩm and 35 kΩm, however, is more frequent in the RES3DINV inversion model.

10.2 Palsa Model

The two inversion models (fig. 10.3) show a slightly higher misfit error (6.38 %) and a higher number of cells (15120) in the RES3DINV inversion model, compared to the BERT inversion model (5.74 %, 13795). The horizontal resistivity distribution reveals that neither the

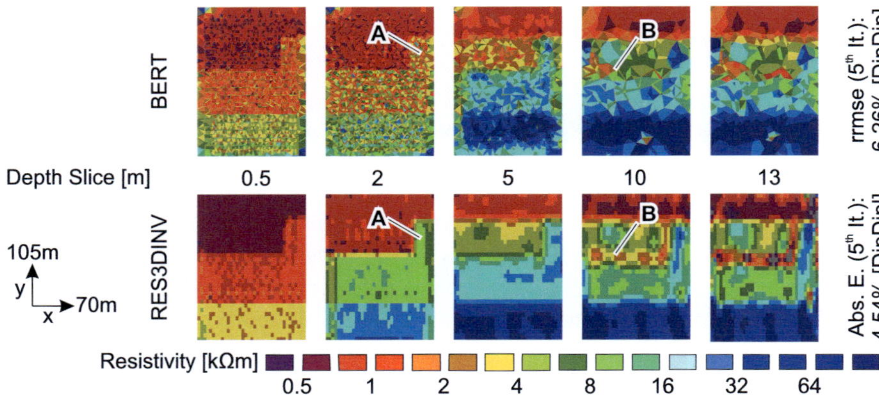

Figure 10.1: Comparison Between Inversion Models (SYN01). a) BERT, b) RES3DINV.

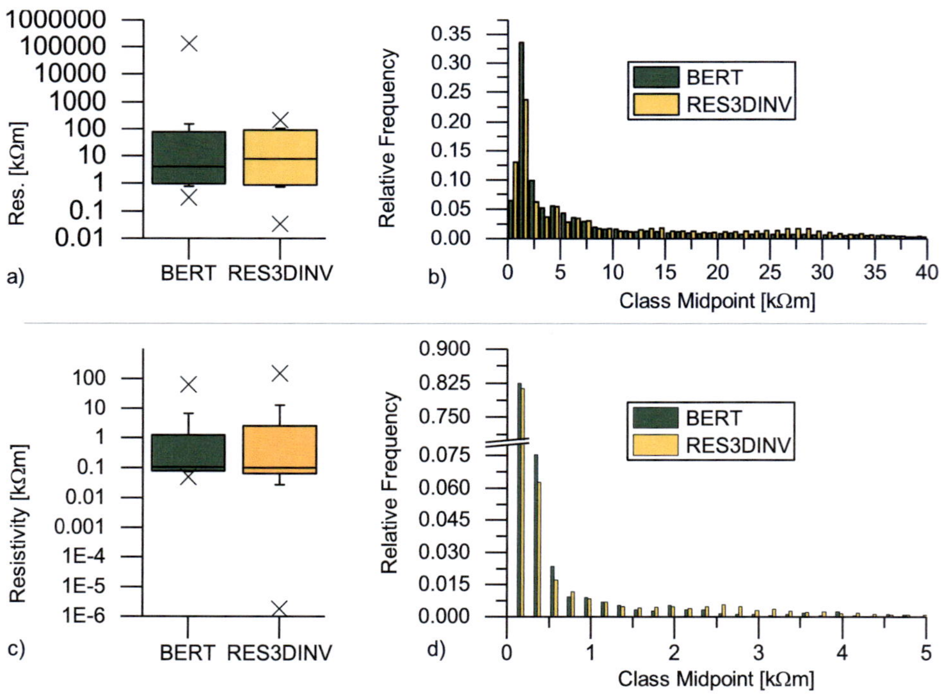

Figure 10.2: Frequency Distribution of Resistivity Values. a) Box Plot SYN01, b) Histogram SYN01, c) Box Plot SYN02, d) Histogram SYN02. Boxes: 25 %/50 %/75 %, Whiskers: 1 %/99 %.

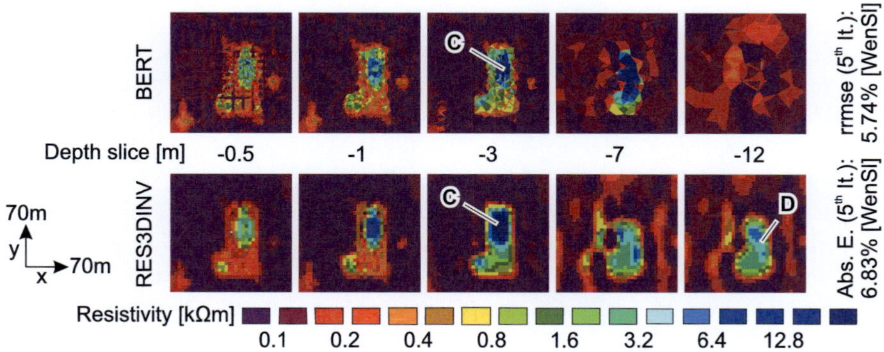

Figure 10.3: Comparison Between Inversion Models (SYN02). a) BERT, b) RES3DINV.

BERT inversion model nor the RES3DINV inversion model reproduces the patchy resistivity distribution pattern of the shallow subsurface of the reference model SYN02 (fig. 3.4b). One prominent structure of relatively high resistivity values appears in the central part of both inversion models (C). The shape of this structure is similar in both inversion models at the depth slices of 0.5 m, 1 m and 2 m and reflects the shape of a corresponding structure in the reference model. Differences between the two inversion models appear in the horizontal depth slice of 7 m, in which the resistivity contrast between structure C and its surrounding area is stronger in the BERT inversion model. The RES3DINV inversion model shows lower resistivity values for structure C than the BERT inversion model and the area which is occupied by structure C is more extensive. Compared to the reference model, the outline of structure C is more accurately reproduced in the BERT inversion model. Another prominent difference between the two inversion models concerns the lowermost presented depth slice of 12 m: only the RES3DINV inversion model, but not the BERT inversion model reproduces the relatively high resistivity values of the reference model (D).

The box plots of the distributions of resistivity values (fig. 10.2c) show that the RES3DINV inversion model comprises a wider range of resistivity values: more than 0.5 % of the resistivity values of the RES3DINV inversion model are below 0.02 kΩm. However, the median of the resistivity values of the BERT inversion model is only slightly higher (0.105 kΩm), than the median of the resistivity values of the RES3DINV inversion model (0.101 kΩm). A similar distribution of resistivity values between the two inversion models is also indicated by the similar frequency distributions of resistivity values (fig. 10.2d): in both inversion models, the class with the highest frequency is the one with the lowermost resistivity values. However, the class of minimum resistivity values has a higher proportion of the total resistivity values in the RES3DINV inversion model than in the BERT inversion model.

10.3 Interpretation

As the options to adjust the inversion parameters differ between the two software products (see sect. 3.1), a detailed analysis of the differences between the inversion models is difficult. The observation that the inversion models of both software products show similar erroneous structures (SYN01: B) and fail to reproduce shallow layers of the reference models, is attributed to the joint step of forward modeling of the reference data. It is therefore assumed that differences between the inversion models and the reference model that appear in both inversion models are not suitable to evaluate inversion quality. Prominent differences that can affect a geomorphological interpretation are absent in the inversion models of the SYN01 data set but appear in the inversion models of the SYN02 data set.

The stronger contrast between structure C and the surrounding area in the depth slice of 7 m in the BERT inversion model of the SYN02 data set indicates stronger interpolation effects in the RES3DINV inversion model. The absence of structure D in the depth slice of 12 m of the BERT inversion model is clearly erroneous and attributed to the low density of data points at this depth level. However, the appearance of structure D in the RES3DINV inversion model is also attributed to undersampling effects: as the shape of this structure differs strongly between the RES3DINV inversion model and the reference model SYN02, it is assumed that it rather originates from a vertical extension of the resistivity pattern of overlaying model layers than from the input data.

Inversion models of both software products contain outliers that are far from the resistivity values in the reference models. Particularly the RES3DINV inversion model of the SYN02 data set shows extremely low resistivity values, which even approach 0 kΩm. Such extremely low resistivity values are absent in the BERT inversion models, but one single outlying data point in the inversion model of the SYN01 data set reaches an unrealistically high resistivity value of more than 100 MΩm. Resistivity values of the BERT inversion models are generally lower than values of the RES3DINV inversion models, as shown by median values and the frequency distributions.

Depth slices of the upper subsurface show more complex resistivity distribution patterns in the BERT inversion models. This is particularly visible in the depth slice of 0.5 m of the inversion model of the SYN02 data set, which reflects the linearly aligned positions of the input data points. The absence of this effect in the corresponding slice of the RES3DINV inversion model affirms the assumption on stronger interpolation effects in the RES3DINV inversion models. These interpolation effects apparently mask indications of undersampling, which are expected in the inversion models of the SYN02 data set, due to the low density of the input data points.

Chapter 11
Discussion

To approach the objectives of this thesis (see sect. 1.4), the presented findings are evaluated in this chapter. They are discussed in the context of recent or landmark literature and compared with results of previously conducted studies. The structure of this chapter is organized in five sections: while the first section evaluates methodological aspects and describes the benefits and drawbacks of the applied geophysical approaches (11.1), the following sections (11.2–11.4) address the different landform types.

11.1 Methodological Aspects

The relatively low misfit errors of the presented ERI models highlight an exceptionally high data quality and successful applications of the different ERI approaches. This is emphasized by the lack of any "w-shaped anomalies" (Kneisel and Hauck, 2008) in plots of the measured values of apparent resistivity (not shown). In total, only 3 % of the measured data points were obviously erroneous and have therefore been removed during processing (see tabs. A.1 to A.4, appendix). Larger adaptions to the targeted line separation of twice the electrode spacing (Gharibi and Bentley, 2005) were unnecessary, as larger obstacles like deep snow fields or huge boulders were absent in the investigated areas on the days of data acquisition. The check of ground resistance values before the start of data acquisition and the eventual attachment of water-soaked sponges to the electrodes, ensured a good ground coupling and limited negative effects on data quality (Hauck and Vonder Mühll, 2003; Marescot et al., 2003).

Although the more robust Wenner-Schlumberger array is commonly recommended for investigations in permafrost environments due to its higher signal strength (Kneisel and Hauck, 2008), surveying with Dipol-Dipol electrode configuration also provided plausible results. When Dipol-Dipol and Wenner-Schlumberger array were used comparatively (not shown), inversion models show comparable resistivity distribution patterns. Despite a superior vertical resolution, the Dipol-Dipol array fails in resolving active layer thicknesses between around 0.3 m and 0.8 m at OVR and in detecting a layer with around 0.3 m thickness in the shallow subsurface of HPA, using an electrode spacing of 2 m. Results of the synthetic data sets show that assumptions on about the uppermost 1 m have to be interpreted with care, even if a high density of measured data points is provided. The ability of ERI to resolve structures of a size that is half the minimum electrode spacing, as described by Sasaki (1992), is affirmed by the inversion models of the SYN02 data set. Problems in detecting the lower limit of permafrost bodies, as reported by Lewkowicz et al. (2011), were avoided by the application of an L1-norm inversion (Loke et al., 2003). The resolution matrix approach, which was used as an indicator for model reliability (see e.g., Stummer et al., 2004), indicates that the upper and central parts of the ERI models are very well resolved in most cases (see fig. B.1, appendix). Low model resolution index values exist only below structures of relatively high resistivity and at the bottoms of the models. These parts of the models were hence interpreted particularly careful.

Differences between intersecting 2-D ERI models (e.g., NAR: E13/E14, OVR: E09/E10) indicate the appearance of "3-D effects" (Chambers et al., 2002; Sjödahl et al., 2006). These effects, which result from small disturbances in complex resistivity distributions, are considered a main drawback of 2-D ERI approach (Sjödahl et al., 2006). Differences between independently inverted 2-D ERI models and corresponding slices of 3-D ERI models of UER01, UER02 and NAR01 are presented and analyzed by Emmert and Kneisel (2017): their study shows that the magnitude of the modeled resistivity values and the depth of the detected structures can vary between the different ERI approaches. The smoothly drawn outlines of the resistivity structures, which can be observed in the 3-D ERI models, are attributed to diagonal filters and interpolations.

Results from the investigations at the high-alpine study sites show relatively low resistivity values compared to other studies from mountainous environments (Dusik et al., 2015; Hilbich et al., 2009; Kneisel, 2010; Maurer and Hauck, 2007). This is attributed to the relatively small grain size at the investigated landform units, as fine-grained materials can store a relatively high amount of liquid water even when temperatures are at 0 °C or even slightly below (Schneider et al., 2013). Relatively low pore space volumes in fine-grained debris are further assumed to limit the formation of larger volumes of ground ice (Scapozza et al., 2011; Vonder Mühll et al., 2000). This is expressed by the lower specific resistivity of frozen fine-grained materials, compared to frozen coarse-grained materials. The observed characteristics of the local resistivity regimes of the high-alpine study sites are presumably transferable to the subarctic study sites, where the occurrence of fine-grained materials is more common (Farbrot et al., 2007; Kneisel et al., 2007; Lewkowicz et al., 2011; Ross et al., 2007).

The detection of massive ground ice in the subsurface of NAR06 by core drilling aligns with the resistivity threshold value stated by Emmert and Kneisel (2017). The depth of the upper boundary of the detected ground ice occurrence in the E24 model corresponds to an increase in resistivity from below 5 kΩm to 8 kΩm. Furthermore, results of this thesis show that resistivity values representing frozen conditions at NAR01 are higher than at the nearby LTF rock glacier (Ikeda and Matsuoka, 2006), but markedly lower than equivalences at other rock glaciers (e.g., Leopold et al., 2011; Ribolini et al., 2010; Seppi et al., 2015). At OVR, the assumed threshold value of 0.6 kΩm to distinguish between frozen and unfrozen subsurface conditions agrees well with observations from other study sites (tab. 11.1).

Results of the comparative analysis between the two different inversion software products show that the selected option to allow the software to assume an unlimited range of resistivity values, is apparently disadvantageous for both software products. However, the negative impact appears to be stronger in the RES3DINV inversion model. Structural differences between the inversion models concern deeper parts with low data coverage. Although corresponding parts of the reference models are better reproduced by RES3DNIV, presented results indicate that this effect is due to interpolation only, which masks an insufficient data coverage. However, it must be noted that the presented analysis is imperfect, as it was not possible to choose comparably adjusted inversion settings.

Data acquisition by GPR is difficult in areas with a blocky surface layer and therefore data acquisition in winter, when antennas can be moved on a continuous snowpack, is usually rec-

Table 11.1: Resistivity Values for Frozen Palsa Materials from Different Studies.

Region	ρ_{min} [kΩm]	Reference
Sweden	1	Sjöberg et al., 2015
Canada	0.3	Lewkowicz et al., 2011
Sweden	0.5	Dobiński, 2010
Canada	1	Fortier et al., 2008

ommended (Maurer and Hauck, 2007). However, surveying at the investigated study sites was performed in summer due to reasons of accessibility. Thus, a weak coupling between the antennas and the ground surface must be taken into account at positions where the antennas could not be placed planar on the surface (Campbell et al., 2018; Heincke et al., 2005; Lambot et al., 2012). To improve ground coupling, antenna orientation was changed from perpendicular broadside to oblique (Baker et al., 2007) at positions where topographical obstacles like boulders disabled a proceeding with parallel antennas. A subsequent disturbance of the electromagnetic field is neglected, as the occurrence of linear objects in the subsurface is not assumed at the investigated study sites (Everett, 2013). The effect of a weak ground coupling is likely visible in the R04 radargram, as the observed weakening in reflection amplitude coincides with a part of the survey line where the antennas were moved along the crack at the surface of OVR05. However, it is also conceivable that the surface crack caused an alteration of the physical properties in the subsurface, which results in an approximation of the dielectric constants and hence a reduction in reflection amplitude.

Compared to other investigations in permafrost environments (e.g., Degenhardt, 2009; Dusik et al., 2015; Monnier and Kinnard, 2015; Sjöberg et al., 2015), the observed penetration depths are relatively low. This is attributed to a relatively high content of liquid water in the subsurface, which causes a strong signal attenuation (Davis and Annan, 1989). While this drawback is often compensated in subarctic environments by a relatively high ice content (e.g., Hinkel et al., 2001), it reduces the suitability of GPR for investigations of landforms that comprise a significant amount of only slightly frozen fine-grained materials. At a bouldery rock glacier, e.g., Monnier et al. (2011) reached a penetration depth of up to 40 m with 50 MHz antennas, while surveying with the same antenna frequency at UER01 resulted in a penetration depth of around 12 m only.

At NAR and LTF, assessments of the vertical EM velocity layering by CMP surveying provided results that agree with the subsurface layering derived from ERI at corresponding positions. This indicates a rather undisturbed layering of the subsurface (Neal, 2004). An estimation of EM velocity values by hyperbola fitting was successfully performed at UER and HPA. Although only relatively small numbers of suitable hyperbolas were detected, likely due to a lack of large boulders in the subsurface, EM velocity values of $0.09\,\mathrm{m\,ns^{-1}}$ and $0.12\,\mathrm{m\,ns^{-1}}$ at UER and $0.08\,\mathrm{m\,ns^{-1}}$ at HPA are in a plausible range (Degenhardt and Giardino, 2003; Dusik et al., 2015; Hubbard et al., 2013; Monnier et al., 2011). Nevertheless, assumed depth levels of the radargrams must be interpreted carefully. At OVR study site, the approach of fitting TWT values to manually sampled values of frost table depth also provided plausible results, compared to other studies from comparable environments (Delisle et al., 2003; Doolittle et al., 1992; Hinkel et al., 2001; Horvath, 1998; Sjöberg et al., 2015). A constant EM velocity value was used for time-depth conversion at FUR, due to the heterogeneous environment and as diffraction hyperbolas were absent. Thus, inaccuracies in the assignment of depth values to subsurface structures must be taken into account and the interpretation of the radargrams is mainly based on characteristics of the reflection patterns and radar facies (Van Overmeeren, 1998).

The separation value of 0.5 m between the parallel survey lines of the 3-D GPR approach performed at OVR exceeds the suggested value for creating a "full-resolution" three-dimensional model (Grasmueck et al., 2005). Nevertheless, this approach is not mandatory to achieve valuable information on the subsurface layering, as proven in various other studies (Godfrey et al., 2008; Munroe et al., 2007). An accurately reproduced frost table geometry indicates the successful application of 3-D GPR at OVR.

11.2 Rock Glaciers

Results from NAR, UER, LTF and partly from FUR show that differences between the investigated rock glaciers concern the ground thermal regime, the formation and distribution of ground ice, frost table characteristics as well as the internal deformation.

The close relationship between the recorded subsurface temperature data and the air temperature values recorded at the weather station in Samedan (MeteoSchweiz, 2018) show a direct coupling between atmospheric and subsurface conditions at NAR01 and UER01. This indicates a high sensitivity of the rock glaciers to microclimatic changes (Schneider et al., 2012). The observed coupling is stronger at NAR01 than it is at UER01, as the insulating effect of a durable snow cover (Rödder and Kneisel, 2012a) is less pronounced at UER01. This is attributed to a higher potential of the furrow-and-ridge topography to accumulate snow at concave surface positions.

A characteristic MAGST distribution pattern for rock glaciers and glacier forefields, with lower values in surface depressions and higher values at rather exposed positions (Hoelzle et al., 1999; Rödder and Kneisel, 2012a), is observed only at the rock glacier snout of UER01. Although this pattern is usually interpreted to reflect differences in the intensity of vertical heat fluxes between permafrost and non-permafrost positions (Gubler et al., 2011; Haeberli and Patzelt, 1982), it is attributed to the formation of an extraordinary durable snow cover in this case. This insulating layer prevents the penetration of warm temperatures in spring and summer. It leads to the relatively low MAGST values at positions that are protected from solar radiation, particularly at the bottom of the deep furrow of the rock glacier snout.

The absence of a characteristic MAGST distribution pattern in the other parts of UER01 is attributed to the pebbly material composition. The presence of rather fine-grained materials affects the regime of energy exchange processes by the ability to store a relatively high amount of liquid water and latent heat (Hoelzle et al., 1999). Cooling mechanisms, which are e.g., caused by air circulations (Hanson and Hoelzle, 2004; Outcalt and Benedict, 1965) are unlikely in pebbly rock glaciers due to the small pore space volumes the subsurface materials. At NAR, relatively high MAGST values in the transition zones between the rock glacier lobes align with the assumed ice-free conditions in the subsurface of these positions.

Results indicate isolated occurrences of congelation ice in the subsurface of the lower lobes of the NAR rock glacier assembly (Haeberli and Vonder Mühll, 1996). This shows that the detected ground ice occurrences are not remnant ice patches from an LIA glaciation, although the corresponding area was marginally affected by surface ice (Coaz et al., 1925; Coaz et al., 1946). However, both the preservation of ground ice and the former occurrence of surface ice are attributed to an extremely low incoming solar radiation (fig. 4.15). The observed small-scale variability in ground ice distribution contradicts the assumption that pebbly materials promote the appearance of balancing energy fluxes by an enhanced advective heat transport due to the higher water retention capacity (Hoelzle et al., 1999; Schneider et al., 2012). This affects not only the lower lobes of the NAR rock glacier assembly, but also the rock glacier snouts of UER01 and FUR02.

Distinctly higher resistivity values appear in the models of the upper lobes of the NAR rock glacier assembly and at LTF rock glacier. Although ancient topographic maps (Coaz et al., 1925; Coaz et al., 1946) indicate the local occurrence of surface ice between 1917 and 1944, resistivity values are still far below the range that is commonly assumed to represent buried glacier ice (Reynolds, 2011). However, it is conceivable that recrystallization processes alter the physical properties of the ground ice, and thereby the electrical conductivity, during multiple freeze-thaw cycles (Reynolds and Paren, 1980). Examples of buried ice of an LIA glaciation in the subsurface of rock glaciers are reported from the French Alps (Monnier et al., 2011), the Central Pyrenees (Lugon et al., 2004) or the Andes (Monnier and Kinnard, 2015). Nevertheless, a direct link between the detected ground ice occurrences at NAR and LTF and buried ice of a former glaciation is speculative.

A different way of ground ice formation is indicated by the longitudinally decreasing ice content of the rock glaciers NAR01 and NAR02. This observation points to the assumption that the origin of the ground ice is refreezing meltwater, which originates from buried patches of surface snow or ice at positions below the talus cones (Haeberli et al., 2006; Haeberli and Vonder Mühll, 1996; Ikeda and Matsuoka, 2006). This concept can be transferred to LTF rock glacier, where it aligns with the particularly intense rockfall activity, indicating that a phase of relatively warm temperatures is present (Gruber and Haeberli, 2007; Matsuoka, 2008).

Ground ice formation by refreezing meltwater from buried patches of snow or ice can further explain the multi-lobe appearance of the rock glaciers at NAR and LTF. This characteristic shape is known to result from a reactivation of single lobes due to an acceleration of lobe movement and is associated with an epigenetic increase in ice content or an enhanced meltwater supply (Müller et al., 2016; Wahrhaftig and Cox, 1959). In combination with the observed spatial variations in grain size distribution, the concept of refreezing meltwater can explain the complex distribution of ground ice in the upper lobes of the NAR rock glacier assembly: Accumulations of larger clasts at the coalescing zones between adjacent talus cones, which are a result of gravitational sorting, lead to a lower water retention capacity at the corresponding positions, which is disadvantageous for refreezing and subsequently impedes the formation of ground ice (Ballantyne and Murton, 2017; Ikeda et al., 2008).

As the observed spatial variations in the grain size distribution at LTF likely affect the characteristics of the local resistivity regimes, a reliable assessment of the ground ice distribution is difficult (Schneider et al., 2013). However, even if the lowermost value for active layer thickness is used, an increase of several meters between the years 2000 and 2016 is observed (Ikeda et al., 2008). This increase in active layer thickness is significantly stronger than increases observed at other rock glacier sites (Kneisel et al., 2014b; Schneider et al., 2012; Zenklusen Mutter and Phillips, 2012). Nevertheless, this is plausible due to site-specific conditions at LTF: the absence of an ice-rich, impermeable frost table and the high content of unfrozen water are assumed to enhance the effect of increasing temperatures (Ikeda et al., 2008; Shur et al., 2005).

A significant influence of snow or ice patches buried by rockfall on the formation of ground ice is negligible at UER01, due to the relatively far distance to the nearest talus slope. Although lateral moraines on the surface of the rock glacier indicate glacier-permafrost interactions (Emmert and Kneisel, 2017), the occurrence of remnant glacier ice in the subsurface of the rock glacier, as e.g., observed by Ribolini et al. (2010) for a study site in the Italian Alps, is excluded by the presented results. Only the absence of frozen conditions in large parts of the root zone of UER01 is attributed to a glacial influence on the distribution of ground ice: it is assumed that the nearby glacier tongue spread on the pre-existing rock glacier during a phase of colder temperatures and thereby disrupted the occurrence of permafrost. Similar concepts are described by Ribolini et al. (2010) and Monnier et al. (2013) for rock glaciers in the Italian and French Alps.

Results indicate that ground ice at UER01 likely originates from refreezing meltwater. This meltwater presumably originates from the surface ice patch, infiltrates into the ground and refreezes at undercooled parts of the subsurface (Humlum et al., 2007). The occurrence of ice of sedimentary origin, formed by recrystallization of snow or ice, is conceivable only at the surface depression in the western part of the rock glacier. The observed decrease in resistivity towards the frontal part of UER01 is attributed to variations in the ionic composition of the ground ice: as the proportion of refreezing meltwater in relation to freezing groundwater decreases with increasing distance from the surface ice patch, the resistivity of the ground ice decreases (Haeberli and Vonder Mühll, 1996; Krainer and Mostler, 2002). The spatial correspondence between the observed structures of relatively high ice content at rock glacier snout and the arcuate surface ridges indicates local enrichments of ground ice due to compressive flow (Haeberli and Vonder Mühll, 1996; Ikeda and Matsuoka, 2006).

The two relatively small rock glaciers NAR07 and UER02 differ from the larger landform units of the rock glacier assembly by a relatively high ice content and a congruent course of frost table and surface topography. The last-named observation is assumed to indicate a "monomorphic" rock glacier type (Frauenfelder and Kääb, 2000), which is set in contrast to the larger rock glaciers for which a successive formation by multiple processes is assumed ("polymorphic rock glacier type"). The topographic position of the two rock glaciers below steep talus slopes indicates a formation by debris-covered avalanche snow deposits (Humlum et al., 2007; Kenner et al., 2017). The importance of this way of ground ice formation on the development of rock glaciers was shown by Kenner and Magnusson (2017), and the concept is also plausible for the development of the other relatively small rock glacier FUR02. This rock glacier is located at a concave surface position where the inclination of the steep slope is reduced to a slope angle below 33°, which is a threshold for the deposition of avalanche snow (Sovilla et al., 2010). Other ways of ground ice formation, e.g., by refreezing meltwater, are unlikely at NAR07 and UER02, regarding the ice-supersaturated subsurface conditions that rather indicate the occurrence of buried snow in the subsurface (Ikeda and Matsuoka, 2006). The assumed existence of an ice lens near the front of FUR02 is attributed to an enhanced preservation of ground ice by active layer thickening due to longitudinal compression (Haeberli and Vonder Mühll, 1996; Scapozza et al., 2011).

The presented ERTM models show that the investigated rock glaciers are less affected by single weather events, like e.g., heavy rainfalls, as other rock glaciers (Hilbich et al., 2011;

Kneisel et al., 2014b). This is attributed to the higher water retention capacity of the pebbly materials, in which the water saturation is generally higher. The observation of an extensive freezing in the upper subsurface of NAR01 and NAR07 shortly after the onset of downward freezing in October illustrates the so-called "autumn-snow effect". This effect, described by Keller (1993), explains an efficient way of ground cooling by the development of only a shallow snow cover in autumn. The effect is caused by the high thermal emissivity of a non-insulating snow cover and the increased surface albedo (Keller and Tamás, 2003). The concept agrees with the recorded GST values and the visual observation of a shallow snow cover on the respective dates of geophysical surveying.

The presented results of geophysical surveying at LTF rock glacier align with the assumed existence of an impermeable frost table, as stated by Ikeda et al., (2008). However, the markedly different results from the nearby rock glaciers NAR01 and NAR07 emphasize the singularity of this phenomenon: these rock glaciers rather suit the traditional concept of an impermeable boundary, as it is described e.g., by Haeberli (1985) and validated by numerous studies (Krainer et al., 2012; Schneider et al., 2013; Vonder Mühll, 1992). This is because only the shape of an impermeable frost table can cause the assumed variations in the intensity of meltwater runoff (Langston et al., 2011; Wright et al., 2009).

At NAR01 and NAR07, the observed rise of the frost table at the beginning of both ERTM investigation periods indicates that meltwater refreezes at a still relatively cold frost table, as described by Humlum et al. (2007) or Sawada et al. (2003) for rock glaciers in Svalbard and Japan, respectively. Together with the simultaneously observed decrease in subsurface temperatures and zero-curtain conditions at the ground surface of NAR01, this indicates oscillating infiltration and warming processes as described by Scherler et al. (2010) and verified by Hilbich et al. (2011). This concept shows that refreezing meltwater can seal the top of the active layer and thereby interrupt heat fluxes from penetrating into the subsurface until a critical water pressure is reached at the ground surface and infiltration continues.

The appearance of undulating reflectors in the radargrams from NAR01, NAR07, LTF rock glacier and FUR02 indicates the occurrence of deformation processes in the respective permafrost layers (Monnier et al., 2011). Spatial coincidences between the appearance of undulating reflectors and resistivity contrasts highlight the influence of ice content variations on movement rates (Kneisel and Kääb, 2007), and the subsequent development of overriding lobes (Barsch, 1996). At LTF, the observation of relatively undisturbed talus deposits in the active layer agrees with the therein detected small movement rates by Ikeda et al. (2008).

Deformation outside the permafrost layers is indicated by the appearance of upwardly dipping reflectors in the radargrams from NAR01 and UER01. The appearance of this reflection pattern affirms a multi-phase rock glacier development of NAR01 and the classification as a polymorphic rock glacier (Frauenfelder and Kääb, 2000). At UER01, where undulating reflectors are absent in the corresponding radargram, it indicates that only the shallow subsurface is affected by contemporary deformation processes.

The appearance of surface-parallel reflectors in the part of the radargram that corresponds to the rock glacier snout of UER01 indicates that this part of the rock glacier was deformed as a whole and that the deformation that caused the development of the furrow-and-ridge topography (Frehner et al., 2015; Kääb and Weber, 2004) was caused rather by external forces. Considering the geomorphologic situation, it is assumed that a glacial override on the upper part of the rock glacier amplified the deformation in the past. This assumption can further explain the observed reflection pattern (Fukui et al., 2008; Krainer et al., 2012). As the two rock glaciers NAR01 and UER01 are located in the back of apparently inactive or relict lobes, respectively, it is assumed that these lobes act or acted as buttresses and additionally enhance or enhanced the deformation processes (Springman et al., 2012).

11.3 Solifluction Lobes

Differences in size between the individual solifluction lobes show that their formation and development are related to small-scale variations of internal or external factors (Matsuoka, 2001). While a map of the potential solar radiation (not shown) reveals that all parts of the slope receive comparable amounts of radiation, an influence of small-scale variations in snow height and snow cover duration are conceivable. Such variations are e.g., known to punctually promote solifluction at positions downward of durable snow patches by an enhanced meltwater supply (Smith, 1988). Although differences in the vegetation pattern between the different lobes were not assessed systematically in the field, a variable density was visually observed during fieldwork. This parameter can control the intensity of diurnal freeze-thaw processes and can thereby enhance solifluction at lobes with rather sparse vegetation (Matsuoka, 2001). Dense mats of *Dryas octopetala*, which is described as an engineering species that closely interacts with solifluction (Eichel et al., 2017), are absent at FUR.

Extraordinary riser heights at some of the solifluction lobes indicate that additional components to solifluction, as e.g., gelifluction or plug-like deformation enhance the rate and intensity of the mass movement. While plug-like deformation only occurs in areas of cold permafrost only (Ballantyne and Murton, 2017), gelifluction, which can form lobes of up to 2 m height (Matsuoka, 2001), is presumably an important component in the formation of the investigated lobes at FUR. This process, however, requires the occurrence of permafrost (Ballantyne and Murton, 2017). Consequently, the observed resistivity values between 8 kΩm and 10 kΩm must represent the lower boundary of resistivity values that indicate frozen conditions. This value aligns with observations from lobe-bearing talus slopes in comparable settings and with similar lobe structures (Otto and Sass, 2006; Scapozza et al., 2011).

The appearance of a small patch with relatively high resistivity values near the front of lobe FUR01 in the E02 model contradicts the observation of Draebing and Eichel (2017), who detected relatively low resistivity values at a corresponding position of a solifluction lobe. However, the resistivity distribution of the E02 model aligns with the spatial variations of volumetric water content measured in the aforementioned study (Draebing and Eichel, 2017). Further resistivity variations in the E02 model are attributed to small-scale differences in the material composition, which e.g., occur when relatively fine-grained materials superimpose materials with a higher amount of larger clasts at the movement base (Benedict, 1970a; Matsuoka, 2010; Matsuoka, 2014).

The occurrence of drainage channels, which Benedict (1970a) assumes to be crucial for the development of lobe-shaped landform units, in contrast to the development of terraces, is indicated by the observation of a linear structure of relatively low resistivity values in the E02 model. The position of such drainage channels may be connected to avalanche tracks, which are important paths for sediment transport and likely occur on the steep talus slope of Blauberg (Freppaz et al., 2010; Schrott et al., 2003). However, it must be noted that the investigated area was snow-free during the days of data acquisition and that the weather was dry on the days before surveying (MeteoSchweiz, 2018). Therefore, meltwater could be provided from ground ice in upward parts of the slope only, and is assumed to percolate through deeper parts of the subsurface (Jaesche et al., 2003).

Considering the permafrost distribution of the entire slope, a non-altitudinal zonation is observed. This aligns with the expected permafrost distribution: The occurrence of permafrost is restricted to lower parts of the slope, while upper parts of the slope are permafrost-free (Kenner et al., 2017; Scapozza et al., 2011). This pattern is attributed to air circulations in coarse-grained talus of deeper subsurface layers (Delaloye and Lambiel, 2005; Lambiel and Pieracci, 2008), or to a locally enhanced thermal insulation by redistributed snow (Lerjen et al., 2003; Luetschg et al., 2004). The occurrence of air circulations in the subsurface is bound to the existence of funnels in the upper and in the lower part of the slope and requires a surface sealing by fine-grained materials in the central part (Harris and Pedersen, 1998; Wakonigg, 1996). Both premises are conceivable at FUR, but due to the lack of GST data or visual inspections in winter, their presence can be neither assured nor neglected. For a comparable talus slope in the Swiss Alps, Scapozza et al. (2011) assumed that ground ice formed as congelation ice at the bottom of the active layer during a past phase of colder climatic conditions, and that it is preserved by enhanced rockfall. Whether this concept is transferable to FUR or not, however, is speculative, due to the lack of additional data.

11.4 Palsas/Lithalsas

The observed correlation between MAGST values and snow cover duration emphasizes the close coupling of these two parameters and the importance of an insulating snow cover on the formation of palsas (Seppälä, 2011). The observed ability of the subsurface materials to hold a relatively high amount of liquid water even in a frozen state aligns with the findings of Kujala et al. (2008), who observed a strong increase in liquid water during a temperature rise from −0.8 °C to 0 °C in peaty materials. However, subsurface materials at OVR differ from common subsurface materials in palsa areas by a relatively low organic content and a high content of sand and thus, transferability of the results is limited (Saemundsson et al., 2012).

The ratio between palsa height and permafrost thickness at palsa OVR01 is around 1/3, which is the characteristic ratio for a palsa that is freely uplifted by buoyancy forces (Allard and Rousseau, 1999; Seppälä and Kujala, 2009). Hence, it is unlikely that the visually observed signs of degradation (i.e., water pond, sunken appearance) can be attributed to a mature stage of palsa development, which starts after the frozen core reached the bottom of the water-saturated layer (Seppälä, 1986). In contrast, these signs of degradation are attributed to bottom-up melting (Seppälä, 2011): this concept describes that the volume of the frozen core starts to shrink due to e.g., changes in water level, climatic changes or small-scale variations in snow cover (Coultish and Lewkowicz, 2003). It lasts until only a small dome-shaped structure that rises out of the water is left.

At OVR02, the discrepancy between the extent of the frozen core and the uplifted area can be explained only if uplift by buoyancy forces is restrained, e.g., due to absence of liquid water below the frozen core (Seppälä and Kujala, 2009). This is usually associated with a mature state of development, after the growing core reached the bottom of the water-saturated layer (Seppälä, 2011). However, a mature state of palsa development disagrees with the assumption that OVR02 is part of the newly formed palsas that developed after the 1960s (Hirakawa, 1986; Kneisel, 2010; Saemundsson et al., 2012). Both observations can be combined, if a compound structure of the frozen core is assumed, as indicated in the R02 radargram: this compound core formed when the growing core of the initial palsa OVR02 reached a pre-existing structure of frozen material at the bottom of the water-saturated layer. If only the thickness of the upper part of this compound core is used for calculating the ratio between palsa height and permafrost thickness, a value is received that represents an initial development stage (Lewkowicz et al., 2011). However, the observation of frozen structures at the bottom of the water-saturated layer is described only rarely in literature. Marklund (2014), who found similar features in a palsa area in northern Sweden, speculates that these structures represent remnant ice of formerly existing, but today degraded palsas.

A deceleration in active layer thickening is indicated by the observation of similar values of active layer thickness at OVR03 and OVR04 between the presented results and investigations in 2010 (Saemundsson et al., 2012). Nevertheless, considering the large size of the palsas, the active layer thickness at OVR03 and OVR04 is relatively low (Seppälä, 2011). This can be explained by surface abrasion due to strong winds on the exposed palsa surface (Seppälä, 2003). At OVR05, the distinct surface crack is presumably responsible for the local thickening of the active layer. However, the formation of cracks on the surface of palsas is not necessarily bound to permafrost degradation, as also dilation due to palsa uplift can cause surface cracking during an early stage of palsa development (Matthews et al., 1997). This assumption is affirmed by the ratio between palsa height and permafrost thickness at OVR05, which, despite ambiguities in assessing permafrost depth from 2-D ERI models, indicates an initial development stage (Allard and Rousseau, 1999; Lewkowicz et al., 2011).

11.5 Patterned Ground

The occurrence of permafrost at HPA aligns with the lower limit of permafrost in Iceland, which is around 800 m a.s.l. (Etzelmüller et al., 2007). However, it contradicts conclusions from MAGST values of the presented GST records and those of the study of Kneisel (2010). This observation shows that conventional concepts that link the occurrence of permafrost to MAGST values, like BTS, are not transferable to the local setting. Such concepts were originally developed for mid-latitudinal mountains and cannot be transferred to the maritime climate of Iceland (Etzelmüller et al., 2007). At HPA01, the presented ERI models align with the previous observations of an 8 m to 10 m thick permafrost body at this position (Kneisel, 2010). The absence of prominent differences between the presented models and results of the previous study shows that subsurface conditions are stable and decoupled from the fluctuating weather conditions. This is considered a sign for a delayed adaption of subsurface conditions to the warming temperatures after the LIA (Etzelmüller et al., 2007). The distinct morphological differences between HPA01 and HPA02 indicate that different processes dominate the formation each occurrence of patterned ground. As both occurrences are separated by a distance of only around 100 m, it is assumed that rather small-scale variations are decisive. This can concern e.g., soil texture, as observed in the field, or pore size, two factors that affect frost-susceptibility and thus the development of patterned ground phenomena (Arnalds, 2015a; Chamberlain, 1981; Hohmann, 1997).

The detection of glacio-fluvial or glacio-lacustrine sediments in the radargrams from HPA01 and HPA02 indicates that meltwater from the nearby glacier leads to the deposition of high amounts of sediments. While strong winds from the glacier erode the deposited fine-grained materials on the plain surface, the lava outcrops that enclose HPA02 act as wind shelters and promote the accumulation of fine-grained materials (Arnalds, 2015b). The occurrence of mudboils (see French, 2018) must be named as another explanation for the accumulation of fine-grained materials at HPA02. However, presented results lack any indications on diagnostic processes like cryostatic pressure, differential frost heaving or soil circulation. Furthermore, the observed diameter of the area covered by fine-grained materials at HPA02 clearly exceeds the range from 1 m to 3 m, which is reported for mudboils (Shilts, 1978).

Results from HPA01 lack any congruence between frost table topography and the appearance of the sorted polygons, as reported from patterned ground occurrences e.g., in Canada (Hubbard et al., 2013) or Svalbard (Kasprzak, 2015). It is therefore assumed that the polygons at HPA01 are not coupled to the recent permafrost occurrence but preserved from a past phase with colder climate. The location of the study site near the margin of Hofsjökull glacier indicates that patterned ground developed in a temporary zone with active periglacial conditions during glacier retreat, as assumed by Haugland (2006) for patterned ground in Norway. Similar to HPA01, presented results from HPA02 lack any indications of a direct coupling between the appearance of patterned ground and characteristics of the internal structure. The small diameter of the polygons and their apparent restriction to a shallow cover layer of fine-grained materials indicate a formation by processes like frost cracking or desiccation (Benedict, 1970b; Washburn, 1979). The formation of desiccation cracks in periglacial environments is reported e.g., from Norway (Ballantyne and Matthews, 1983; Matthews et al., 1998) or Svalbard (Van Vliet-Lanoë, 1988). It requires an occasional water-saturation and strong evaporation, which is both conceivable at HPA (Gutiérrez and Gutiérrez, 2016).

Chapter 12
Conclusions

The final chapter summarizes the key findings (12.1) and replies on the research questions stated at the beginning of this thesis (12.2). Furthermore, it evaluates the importance of the presented findings and includes an outlook on possible future research (12.3).

12.1 Key Findings

Findings illustrate the potential of the internal structure to archive past environmental conditions and processes. This provides valuable information for an enhanced understanding of the development of periglacial landforms. This is important for the interpretation of results from numerical models that e.g., describes rock glacier dynamics, as such models only rarely include information on subsurface heterogeneities or melting processes of internal ice (Haeberli et al., 2006; Müller et al., 2016).

At several rock glaciers of the NAR rock glacier assembly and at UER01, the detected internal features indicate a discrepancy between the recent environmental conditions and the size and appearance of the landform units. A similar discrepancy is indicated between the visible polygons and the recent distribution of permafrost at HPA01. The internal structure of the palsas at OVR reveals a long-lasting palsa development in the investigated area, that recently faces environmental changes. Rather uniform subsurface conditions at the smaller landform units NAR07, UER02, HPA02, FUR02 indicate a formation by one distinct formative process. Intraannual variations in the internal structure of NAR01 and NAR07 reflect short-term processes like meltwater flow and complete the scope of a multi-temporal archive.

Findings highlight that the incorporation of surface snow or ice into the subsurface is crucial for the development of many landform units in high-alpine environments. Ground ice of this epigenetic origin can affect or even initiate subsequent processes, like e.g., reactivate formerly inactive rock glacier lobes or accelerate rock glacier movement by an enhanced supply of meltwater. Results from UER show that glacier-permafrost interactions caused a disruption of permafrost conditions in the subsurface of a rock glacier.

The detection of isolated patches of ground ice in unfrozen surroundings shows that the influence of balancing heat fluxes, which is commonly assumed to be strong in fine-grained materials close to the freezing point, is relatively low at the investigated study sites. This indicates a relatively low sensitivity of the investigated landform units to environmental changes and may lead to an underestimation of the potential of pebbly materials to bear ground ice. Results show that the actual extent of permafrost exceeds the extent of the potential permafrost area as depicted by the APIM or the Swiss Map of Potential Permafrost Distribution (BAFU, 2005; Boeckli et al., 2012). Findings of this work show examples for the concepts of oscillating infiltration into frozen subsurface materials (Scherler et al., 2010) and the "autumn-snow-effect" (Keller, 1993). They are further assumed to display the concept of (iii) bottom-up melting (Seppälä, 1986).

In this thesis, characteristic resistivity values and characteristic reflection patterns, as described in prior studies, are used for interpreting the ERI models and radargrams, respectively. This approach permits a differentiation between the landform types, which is particularly shown by results from FUR, where the detection of undulating reflectors enables to distinguish between solifluction lobes and pebbly rock glaciers. The detection of small-scale subsurface variations shows that three-dimensional geophysical approaches are superior to two-dimensional approaches in complex environments, as small structures could be missed by small-scale probing or point sampling. This can cause severe misinterpretations when subsurface parameters are extrapolated. Based on the presented results, the research questions stated in chapter 1.4 can be answered as follows:

Group 1: Composition of the Investigated Landform Units

Are the investigated study sites/landform units currently affected by permafrost conditions?

Results show current permafrost conditions at all study sites. However, the influence of permafrost on landform formation and development differs between the investigated landform units: Current permafrost conditions are assumed to affect the development of the upper rock glacier lobes of NAR as well as UER02, FUR02, LTF rock glacier and the palsas of OVR. In contrast, the isolated occurrences of permafrost in the subsurface of UER01 are presumably no longer involved in the development of the rock glacier. A similar situation is assumed at HPA, where the widespread occurrence of permafrost is likely decoupled from the development of the patterned ground phenomena. At FUR, permafrost currently affects only deeper parts of the talus slope and is not considered a crucial factor for the development of the solifluction lobes.

To what extent do spatial variations of the internal structure (e.g., active layer thickness, frost table topography, ice content) occur?

Differences in the internal structure appear between landform types and between landform units. The investigated rock glaciers show strong variations in ice content, ranging from ice-rich (NAR07, UER02) to only isolated occurrences (UER01). Where a continuous frost table exists, it can be impermeable (NAR01) or permeable (LTF). The shape of the frost table ranges from surface parallel (NAR07) or undulating (NAR01). Active layer thicknesses are also highly variable: frozen conditions can exist directly below the surface (LTF) or only at greater depths (UER01). The internal structure of the investigated solifluction lobes lacks any permafrost-related features, as permafrost conditions affect only deeper parts of the slope below the lobes (FUR01). Strong ice content variations appear between the investigated palsas at OVR and are attributed to different development stages: barely frozen conditions are assumed at a degrading palsa (OVR01), while palsas of a mature state comprise a relatively high amount of ice (OVR03). Active layer thickness variations are less pronounced at the palsas, compared to the rock glaciers, likely because the frozen cores are in an equilibrium of buoyancy with the surrounding water-saturated layer. A deviation from the characteristic dome-shaped frost table topography is observed at the palsa with the surface crack (OVR05).

Smaller variations in ice content appear between the two areas of patterned ground (HPA01/HPA02), and results show a rather flat frost table topography at both spots.

What is the origin of the detected ground ice?

Results indicate that altered ice of sedimentary origin occurs below the talus cones of NAR and LTF as well as at the surface depression in the central part of UER01. At these positions, ground ice is assumed to originate from incorporated surface patches of snow or ice and is likely affected by alternating melting and refreezing. Refreezing meltwater from these ground ice occurrences is subsequently assumed to be the origin of the ground ice at the upper rock glacier lobes of NAR. The occurrence of ground ice in the root zone of UER01 presumably originates from refreezing meltwater of the surface ice patch. At NAR07, UER02 and FUR02, a ground ice formation from debris covered avalanche snow deposits is assumed. The origin of the ground ice at OVR is evident, as the formation of palsas is by definition related to the occurrence of segregation ice. Results show that the occurrence of sedimentary ice can be excluded at HPA, and it is therefore assumed that the formation of ground ice presumably started only after, but at the earliest during the retreat of the glacier.

Which prominent resistivity structures or GPR reflection patterns appear in the results of different landform units or landform types?

Structures of relatively high resistivity values, interpreted as areas with ice-saturated conditions, recur in the root zones of multiple investigated rock glaciers. As their appearance coincides with chaotic patterns of short reflectors at corresponding positions in overlapping radargrams, they clearly indicate the absence of massive ice. Indications for the occurrence of massive ice are e.g., blank areas without any reflectors, as they can be observed in the radargrams from the investigated palsas. Here, they coincide with structures of relatively high resistivity values in the corresponding ERI models and outline the extents of the frozen palsa cores. While radargrams from the smaller rock glaciers comprise undulating reflectors as a common feature, upwardly dipping reflectors solely occur in radargrams from the larger rock glaciers. This highlights the occurrence different types of deformation and aligns with the observation of markedly higher resistivity values in the ERI models from the smaller rock glaciers. The range of these values indicates stronger differences in ice content and a less advanced stage of ice recrystallization.

Radargrams from the smaller rock glaciers and from the investigated solifluction lobe FUR01 show a characteristic reflection pattern for stratified talus deposits in the shallow subsurface. The ERI model from the complete talus slope at FUR shows patches of relatively high resistivity values below the risers of the solifluction lobes. This observation is attributed to small-scale differences in the material composition and subsequent differences in the water retention capacity.

Does the spatial distribution of ground ice correspond to the spatial distribution of surface parameters, such as surface topography or vegetation?

At NAR, the distribution of ground ice reflects the outlines of the rock glaciers, while transitional areas between the lobes are ice-free. The occurrence or absence of permafrost at

UER01 is connected to different positions of the furrow-and-ridge topography. The outlines of the solifluction lobes at FUR correlates with the resistivity distribution in the upper part of the subsurface, but as the lobes are apparently not directly affected by permafrost conditions, this correspondence is attributed to differences in the surface material composition. The segmentation of the investigated area at LTF into parts with a different grain size, is also reflected in the resistivity distribution. At OVR, the observed vegetation patterns correlate with the uplifted area, but as the vegetation pattern likely follows an increase in soil drainage, this connection not directly involves the subsurface. The higher water retention capacity of fine-grained materials is presumably responsible for the correspondence between grain size and resistivity distribution at HPA.

Group 2: Permafrost and Landform Development

How does the internal structure interact with geomorphologic processes?

At the upper lobes of NAR, as well as at UER02, LTF and FUR, intense rockfall or avalanche activity is presumably responsible for the formation and distribution of ground ice. Closely connected to the distribution of ground ice is an enhanced meltwater flow, which affects positions in flow direction. Refreezing meltwater can lead to a redistribution of ground ice and cause a gradually decrease in ice content, as it is absorbed by the subsurface materials, depending on their retention capacity. At NAR, it is assumed that this process chain is responsible for a partial reactivation of rock glacier lobes. Seasonal variations in the meltwater flow at NAR01 and NAR07 are apparently affected by the shape of the frost table. It is further assumed that isolated ice patches in the subsurface of NAR06 and NAR04 act as heat sinks that punctually absorb energy and distribute it over a more extensive area. The occurrence of ground ice in the subsurface of the talus slope at FUR facilitates mass wasting processes, specifically gelifluction, by forming a sliding plane. Furthermore, it is a source of meltwater, which both promotes solifluction by an enhanced availability of water. At UER01, deformation processes apparently control the enrichment of ground ice at the rock glacier snout. Differences in ice content and permafrost thickness between the palsas at OVR indicate different stages of palsa development and thus the dominance of different processes. This includes the long-term or short-term aggregation of ice, e.g., at OVR03 or OVR02, respectively, as well as the assumed bottom-up melting at OVR01. At HPA, the occurrences of patterned ground are presumably decoupled from the occurrence of permafrost, but still show interactions between the internal structure and the process domain: The formation of surface cracks at HPA02 is likely promoted by a layer of fine-grained materials in the upper part of the subsurface.

Which processes determined landform development in past phases of colder climate?

It is assumed that a lower rockfall activity has affected landform development at NAR, LTF and FUR by a reduced debris supply. A more widespread distribution of surface ice, as depicted on ancient topographic maps at NAR and LTF, would have had disrupted the occurrence of permafrost, although its depicted extent does not allow any statement about the thickness of the surface ice. Nevertheless, the dimension and the multi-lobe appearance of the rock glaciers indicate that periglacial processes prevailed in the past. Assuming ice-free surface conditions at FUR, a more widespread permafrost distribution is assumed to have caused an enhanced gelifluction component, and that resulting lobes were therefore higher. At UER01, it is assumed that the adjacent glacier tongue was overriding the rock glacier during past phases of colder climate. This presumably enhanced the internal deformation of the rock glacier through immense external thrusting forces and initiated the development of the furrow-and-ridge topography. It is assumed that such override events caused a disruption of permafrost in the subsurface. As results of this thesis indicate a long-term development of palsas at OVR, it is assumed that climatic changes in the past may have affected the intensity of palsa formation, but not fundamentally changed the process regime. Colder climatic conditions are assumed to have contributed to the formation of patterned ground at HPA01, although a formation of this surface phenomenon does not generally require the occurrence of permafrost. However, the development of patterned ground at HPA can have started only after the begin of glacier retreat. During past phases of colder climate, HPA was likely covered by ice and formed by glacial processes.

How will warming temperatures alter the permafrost distribution and affect the landform development?

Although warming air temperatures will increase heat fluxes into the subsurface, results of this thesis indicate that even fine-grained materials have a strong ability to buffer this increase. This is illustrated by subsurface temperature records from NAR and UER, which show stable or decreasing temperatures throughout the investigated time span. At the landform units that are connected to intense rockfall activity, warming temperatures will enhance debris supply and subsequently promote the incorporation of surface snow or ice into the subsurface. This may result in a reactivation of formerly inactive parts of the landform units, which, however, will only last as long as there is a considerable amount of snow. A loss of ground ice at the high-alpine study sites will result in enhanced surface cracking, as already indicated at NAR04 and NAR06, and subsequently in a collapse of affected rock glacier lobes. Variations in snowfall are hardly assessable, but, as indicated by the presented GST records, will have a strong impact on the insulation of the subsurface. Furthermore, temporal variations in snowfall will affect the strength of the "autumn-snow effect", which is assumed to be an important factor for cooling the subsurface. At OVR, warming temperatures are assumed to enhance palsa degradation and bottom-up melting will likely affect a higher number of palsas. Variations in the distribution of snow will strongly affect the formation of new palsas, as these processes are closely related. The occurrence of patterned ground at HPA01 is apparently not directly influenced by permafrost anymore, and it is hence assumed to vanish as the polygonal structures will be filled by deposited fine-grained materials. Nevertheless, an enhanced glacier retreat will promote the formation of patterned ground at positions that will become ice-free.

Group 3: Methodological Approach

Are the applied methods suitable to answer the research questions?

The applied geophysical methods are well-established for investigations in periglacial areas. However, the presented findings show that the availability of ground truth information is sometimes important for an unambiguous interpretation of the results. This concerns areas where a relatively low ground ice content or very complex subsurface conditions are assumed, like FUR or LTF. Although results permit a general assessment of the internal structure at these two study sites, a detailed interpretation of the detected structures is difficult. While ERI provides information on the existence of ground ice and its characteristics, GPR provides information on the structural layering of the subsurface. The joint application of the two methods thus permits a differentiation between different types of ground ice (UER01) or between solifluction lobes and pebbly rock glaciers (FUR). The ability of joint applications to overcome weaknesses of a single method is shown e.g., at OVR03, where ERI fails in resolving the thickness of the active layer and at all sites with fine-grained debris, where the penetration depth of GPR is strongly reduced due to a relatively high content of liquid water. However, it must be noted that ERI and GPR are both based on the dielectric properties of the subsurface materials and are hence not fully independent.

Records of subsurface temperature data provide ground truth information on the existence of permafrost and permit to deduce a threshold value to distinguish between frozen and unfrozen subsurface conditions. However, the relatively low accuracy of the subsurface temperature sensors at NAR and UER prevents unambiguous assessments of the conditions in the subsurface. GST records are assumed to be of secondary importance for interpreting the results from sites with fine-grained materials, as the ground thermal regime in such areas presumably differs from the ground thermal regime in other permafrost areas. Recorded pF-values show plausible results at OVR and enable the detection of year-round water-saturated conditions at a position between the palsas. However, the approach has still to prove its value in more complex settings.

Which advantages provides the relatively innovative and until today sparely used 3-D ERI approach over the established 2-D ERI approaches? Which limitations exist and how can the 3-D ERI approach be further developed?

Key advantage of 3-D ERI approaches is the ability to perform spatial analyses between the horizontal resistivity distribution and other horizontally distributed parameters. This allows to perform statistical analyses and can thus help to investigate relationships between surface and subsurface parameters, like e.g., between ice content and vegetation. However, as the focus of this thesis was to assess internal characteristics, congruencies between surface and subsurface parameters were investigated only qualitatively. Furthermore, it is assumed that results of 3-D ERI better reflect the real subsurface conditions, compared to results of 2-D ERI, as "3-D effects" are absent. Hence, a detection of small-scale changes in subsurface conditions and the delimitation of isolated structures is possible. Limitations of 3-D ERI are a lower flexibility in the survey setup, as obstacles like large boulders or surface ice patches can be hardly bypassed and require adjustments in line separation. However, this was a minor problem at the investigated sites and a sufficient data coverage was likely achieved in all inversion models. Another disadvantage of 3-D ERI is the generally higher time consumption of both survey setup and data inversion. The use of tape measures to determine start and end positions of the two-dimensional survey lines for q-3-D data acquisition is considered an efficient procedure to avoid problems due to the misalignment of electrodes. In an exemplary study with data from UER02 (not shown), the option to directly include RTK-GNNS positioning data into the inversion scheme provided only small improvements in misfit error over the traditional approach outlined in section 3.1. Data acquisition by the real-3-D approach is reasonable only for very small-scale investigations, as cable length and the number of electrodes increase exponentially.

Which differences occur between resistivity models generated with different inversion software?

The comparative analysis between the two inversion models, generated with the software products RES3DINV and BERT, respectively, shows only slight differences. Both software products generate results that would lead to a similar geomorphological interpretation. Observed differences in lower parts of the inversion models show that areas with low data coverage must be interpreted carefully. Hence, the execution of approaches that enable assessments of model reliability, such as e.g., DOI or model resolution, is strongly advised. It is concluded that the freely licensed, open source software product BERT is a valuable tool and can substitute the widespread RES2D/3DINV software package.

12.2 Outlook

This thesis can be a base for further research on different objectives. For a better understanding of the interactions between surface and subsurface conditions, future studies should include a higher number of horizontally distributed surface parameters like temperature, soil texture or vegetation. Through the application of 3-D ERI, these parameters can be linked to the horizontal resistivity distribution in the subsurface, which reflects not only the distribution of ground ice, but also e.g., the distribution of soil moisture. Existing studies using three-dimensional ERI approaches (see sect. 1.3) show their potential for resistivity mapping especially in complex environments, as the appearance of "3-D effects" is avoided. From a methodological view, future research should comprise three-dimensional geoelectrical monitoring surveys, as this approach can provide information on parameters which are variable in space and time, as e.g., meltwater flow.

To validate assumptions on the internal structure, future research should include more and different complementary methods. This comprises other geophysical techniques, such as SRT, but also a transfer of techniques from neighboring fields, such as thermal infrared surveying (see e.g., Stilla and Hoegner 2007). Especially at study sites with a shallow active layer thickness, the last-named approach can provide valuable information on the ground thermal regimes. Longer time-series of recorded temperature data or pF-values permits the detection of long-term adaptions of the landform units to climate change. As results from NAR show that ground truth information can be also achieved by low-cost core drilling, this approach should be used more extensively in future research. Furthermore, the promising results of this thesis advise the transfer of the presented approach to other landform types, such as bouldery rock glaciers, pingos or different types of patterned ground.

Combined with the presented assessments of the internal structure, investigations of landform dynamics can affirm e.g., the assumed movement of the upper lobe of NAR07. Investigations of landform dynamics can further enhance the understanding of landform development at all study sites. Data can be gained e.g., by photogrammetric techniques or by terrestrial laser scanning. Especially the repeated application of Structure-from-Motion techniques, in combination with UAV-based surveying, permits large-scale investigations. This approach can e.g., provide valuable information on the formation of new palsas at OVR. The observation of different resistivity values representing the lower limit of frozen conditions between the study sites should be analyzed systematically. A collection of these threshold values from multiple areas can provide valuable information for investigations at sites where an assessment of the resistivity-permafrost relationship is not possible.

Bibliography

Åkerman, H.J., Johansson, M., 2008. Thawing permafrost and thicker active layers in sub-arctic Sweden. Permafrost and Periglacial Processes, 19(3), 279–292.

Allard, M., Rousseau, L., 1999. The international structure of a palsa and a peat plateau in the Rivière Boniface region, Québec: Interferences on the formation of ice segregation mounds. Géographie physique et Quaternaire, 53(3), 373–387.

Annan, A., 2009. Electromagnetic principles of ground penetrating radar. In: H.M. Jol (Ed.), Ground Penetrating Radar Theory and Applications Elsevier, Amsterdam, pp. 1–40

Annan, A., Cosway, S., 1992. Ground penetrating radar survey design, Symposium on the Application of Geophysics to Engineering and Environmental Problems 1992. Society of Exploration Geophysicists, pp. 329–351.

Annan, A., Davis, J., 1976. Impulse radar sounding in permafrost. Radio Science, 11(4), 383–394.

Archie, G.E., 1942. The electrical resistivity log as an aid in determining some reservoir characteristics. Transactions of the AIME, 146(01), 54–62.

Arnalds, Ó., 2010. Dust sources and deposition of aeolian materials in Iceland. Icelandic agricultural sciences, 23, 3–21.

Arnalds, Ó., 2015a. Frost and the Soil Environment, The Soils of Iceland. Springer Netherlands, Dordrecht, pp. 119–137.

Arnalds, Ó., 2015b. The Volcanic Aeolian Environments of Iceland, The Soils of Iceland. Springer Netherlands, Dordrecht, pp. 139–152.

Auer, I., Böhm, R., Jurkovic, A., Lipa, W., 2007. HISTALP-historical instrumental climatological surface time series of the Greater Alpine Region. International Journal of Climatology, 27(1), 17–46.

Azócar, G., Brenning, A., 2010. Hydrological and geomorphological significance of rock glaciers in the dry Andes, Chile (27–33 S). Permafrost and Periglacial Processes, 21(1), 42–53.

BAFU, 2005. Hinweiskarte der potenziellen Permafrostverbreitung. Available from: https://map.geo.admin.ch (last accessed: 22.01.2019).

Baker, G.S., Jordan, T.E., Pardy, J., 2007. An introduction to ground penetrating radar (GPR). Special Papers - Geological Society of America, 432, 1–18.

Ballantyne, C.K., Murton, J.B., 2017. Periglacial Geomorphology. John Wiley & Sons, Somerset, 888 pp.

Ballantyne, C.K., Matthews, J.A., 1983. Desiccation cracking and sorted polygon development, Jotunheimen, Norway. Arctic and Alpine Research, 15(3), 339–349.

Barker, R., Moore, J., 1998. The application of time-lapse electrical tomography in groundwater studies. The Leading Edge, 17(10), 1454–1458.

Barsch, D., 1993. Periglacial geomorphology in the 21st century. Geomorphology, 7(1), 141–163.

Barsch, D., 1996. Rockglaciers: indicators for the present and former geoecology in high mountain environments. Springer, Berlin, 319 pp.

Bearth, P., Heierli, H., Roesli, F., Furrer, H., 1987. Albulapass. Geologischer Atlas der Schweiz 1:25 000. Available from: https://map.geo.admin.ch (last accessed: 22.01.2019).

Benedict, J.B., 1970a. Downslope soil movement in a Colorado alpine region: rates, processes, and climatic significance. Arctic and Alpine Research, 2(3), 165–226.

Benedict, J.B., 1970b. Frost cracking in the Colorado Front Range. Geografiska Annaler: Series A, Physical Geography, 52(2), 87–93.

Benedict, J.B., 1976. Frost creep and gelifluction features: a review. Quaternary Research, 6(1), 55–76.

Beniston, M., Uhlmann, B., Goyette, S., Lopez-Moreno, J.I., 2011. Will snow-abundant winters still exist in the Swiss Alps in an enhanced greenhouse climate? International Journal of Climatology, 31(9), 1257–1263.

Bentley, L.R., Gharibi, M., 2004. Two-and three-dimensional electrical resistivity imaging at a heterogeneous remediation site. Geophysics, 69(3), 674–680.

Berthling, I., Etzelmüller, B., 2011. The concept of cryo-conditioning in landscape evolution. Quaternary Research, 75(2), 378–384.

Berthling, I., Etzelmüller, B., Isaksen, K., Sollid, J.L., 2000. Rock glaciers on Prins Karls Forland. II: GPR soundings and the development of internal structures. Permafrost and Periglacial Processes, 11(4), 357–369.

Berthling, I., Melvold, K., 2008. Ground-penetrating radar. In: C. Hauck, C. Kneisel (Eds.), Applied Geophysics in Periglacial Environments. Cambridge University Press, Cambridge, pp. 81–97.

Binley, A., 2015. Tools and Techniques: Electrical Methods. In: G. Schubert (Ed.), Treatise on Geophysics (Second Edition). Elsevier, Oxford, pp. 233–259.

Blume, H.-P., Brümmer, G.W., Fleige, H., Horn, R., Kandeler, E., Kögel-Knabner, I., Kretzschmar, R., Stahr, K., Wilke, B.-M., 2016. Scheffer/Schachtschabel Soil Science. Springer, Berlin, Heidelberg, 618 pp.

Bodin, X., Rojas, F., Brenning, A., 2010. Status and evolution of the cryosphere in the Andes of Santiago (Chile, 33.5°S.). Geomorphology, 118(3–4), 453–464.

Boeckli, L., Brenning, A., Gruber, S., Noetzli, J., 2012. A statistical approach to modelling permafrost distribution in the European Alps or similar mountain ranges. The Cryosphere, 6(1), 125–140.

Bolch, T., Gorbunov, A.P., 2014. Characteristics and origin of rock glaciers in Northern Tien Shan (Kazakhstan/Kyrgyzstan). Permafrost and Periglacial Processes, 25(4), 320–332.

Bolch, T., Rohrbach, N., Kutuzov, S., Robson, B., Osmonov, A., 2019. Occurrence, evolution and ice content of ice-debris complexes in the Ak-Shiirak, Central Tien Shan revealed by geophysical and remotely-sensed investigations. Earth Surface Processes and Landforms, 44(1), 129–143.

Borge, A.F., Westermann, S., Solheim, I., Etzelmüller, B., 2017. Strong degradation of palsas and peat plateaus in northern Norway during the last 60 years. The Cryosphere, 11(1), 1–16.

Brandt, O., Langley, K., Kohler, J., Hamran, S.-E., 2007. Detection of buried ice and sediment layers in permafrost using multi-frequency Ground Penetrating Radar: A case examination on Svalbard. Remote Sensing of Environment, 111(2-3), 212–227.

Briggs, M.A., Campbell, S., Nolan, J., Walvoord, M.A., Ntarlagiannis, D., Day-Lewis, F.D., Lane, J.W., 2017. Surface geophysical methods for characterising frozen ground in transitional permafrost landscapes. Permafrost and Periglacial Processes, 28(1), 52–65.

Bristow, C.S., 2009. Ground penetrating radar in aeolian dune sands. In: H.M. Jol (Ed.), Ground Penetrating Radar Theory and Applications, Elsevier, Amsterdam, 273–297.

Burga, C.A., Frauenfelder, R., Ruffet, J., Hoelzle, M., Kääb, A., 2004. Vegetation on Alpine rock glacier surfaces: a contribution to abundance and dynamics on extreme plant habitats. Flora - Morphology, Distribution, Functional Ecology of Plants, 199(6), 505–515.

Campbell, S., Affleck, R.T., Sinclair, S., 2018. Ground-penetrating radar studies of permafrost, periglacial, and near-surface geology at McMurdo Station, Antarctica. Cold Regions Science and Technology, 148, 38–49.

Cassidy, N.J., 2009. Ground penetrating radar data processing, modelling and analysis. In: H. Jol (Ed.), Ground penetrating radar: theory and applications, Elsevier, Amsterdam, pp. 141–176.

Chamberlain, E.J., 1981. Frost susceptibility of soil, review of index tests, Cold Regions Research and Engineering Lab Hanover NH. CRREL Monograph 81-2, 124 pp.

Chambers, J., Ogilvy, R., Kuras, O., Cripps, J., Meldrum, P., 2002. 3D electrical imaging of known targets at a controlled environmental test site. Environmental Geology, 41(6), 690–704.

Coaz, J.W.F., Leuzinger, R., 1878. Bevers, Topographischer Atlas der Schweiz Eidg. Landestopographie, Bern. Available from: https://map.geo.admin.ch (last accessed: 22.01.2019).

Coaz, J.W.F., Leuzinger, R., Held, L., 1925. St. Moritz, Topographischer Atlas der Schweiz. Eidg. Landestopographie, Bern. Available from: https://map.geo.admin.ch (last accessed: 22.01.2019).

Coaz, J.W.F., Leuzinger, R., Held, L., 1946. St. Moritz, Topographischer Atlas der Schweiz. Eidg. Landestopographie, Bern. Available from: https://map.geo.admin.ch (last accessed: 22.01.2019).

Cockett, R., Kang, S., Heagy, L.J., Pidlisecky, A., Oldenburg, D.W., 2015. SimPEG: An open source framework for simulation and gradient based parameter estimation in geophysical applications. Computers & Geosciences, 85, Part A, 142–154.

Coultish, T., Lewkowicz, A., 2003. Palsa dynamics in a subarctic mountainous environment, Wolf Creek, Yukon Territory, Canada, Proceedings of the 8th International Conference on Permafrost, Zurich, Switzerland.

Croce, F.A., Milana, J.P., 2002. Internal structure and behaviour of a rock glacier in the Arid Andes of Argentina. Permafrost and Periglacial Processes, 13(4), 289–299.

Dahlin, T., Zhou, B., 2004. A numerical comparison of 2D resistivity imaging with 10 electrode arrays. Geophysical prospecting, 52(5), 379–398.

Davis, J.L., Annan, A.P., 1989. Ground-penetrating radar for high-resolution mapping of soil and rock stratigraphy. Geophysical prospecting, 37(5), 531–551.

Degenhardt, J.J., 2009. Development of tongue-shaped and multilobate rock glaciers in alpine environments–Interpretations from ground penetrating radar surveys. Geomorphology, 109(3-4), 94–107.

Degenhardt, J.J., Giardino, J.R., 2003. Subsurface investigation of a rock glacier using ground-penetrating radar: Implications for locating stored water on Mars. Journal of Geophysical Research: Planets, 108(E4).

Delaloye, R., Lambiel, C., 2005. Evidence of winter ascending air circulation throughout talus slopes and rock glaciers situated in the lower belt of alpine discontinuous permafrost (Swiss Alps). Norsk Geografisk Tidsskrift-Norwegian Journal of Geography, 59(2), 194–203.

Delaloye, R., Perruchoud, E., Avian, M., Kaufmann, V., Bodin, X., Hausmann, H., Ikeda, A., Kääb, A., Kellerer-Pirklbauer, A., Krainer, K., 2008. Recent interannual variations of rock glacier creep in the European Alps, Proceedings of the 9th International Conference on Permafrost, Fairbanks, Alaska. Citeseer, pp. 343–348.

Delisle, G., Allard, M., Fortier, R., Calmels, F., Larrivée, É., 2003. Umiujaq, Northern Québec: innovative techniques to monitor the decay of a lithalsa in response to climate change. Permafrost and Periglacial processes, 14(4), 375–385.

Dobiński, W., 2010. Geophysical characteristics of permafrost in the Abisko area, northern Sweden. Polish Polar Research, 31(2), 141–158.

Dobiński, W., 2011. Permafrost. Earth-Science Reviews, 108(3), 158–169.

Doolittle, J., Hardisky, M., Black, S., 1992. A ground-penetrating radar study of Goodream palsas, Newfoundland, Canada. Arctic and Alpine Research, 24(2), 173–178.

Draebing, D., 2016. Application of refraction seismics in alpine permafrost studies: A review. Earth-Science Reviews, 155, 136–152.

Draebing, D., Eichel, J., 2017. Spatial Controls of Turf-Banked Solifluction Lobes and Their Role for Paraglacial Adjustment in Glacier Forelands. Permafrost and Periglacial Processes, 28(2), 446–459.

Dusik, J.M., Leopold, M., Heckmann, T., Haas, F., Hilger, L., Morche, D., Neugirg, F., Becht, M., 2015. Influence of glacier advance on the development of the multipart Riffeltal rock glacier, Central Austrian Alps. Earth Surface Processes and Landforms, 40(7), 965–980.

ecoTech, 2014. TensioMark - Manual for installation and use of Tensiomarks. Available from: https://www.stevenswater.com/resources/documentation/Stevens_Tensiomark_Manual.pdf (last accessed: 22.01.2019).

Eichel, J., Corenblit, D., Dikau, R., 2016. Conditions for feedbacks between geomorphic and vegetation dynamics on lateral moraine slopes: a biogeomorphic feedback window. Earth Surface Processes and Landforms, 41(3), 406–419.

Eichel, J., Draebing, D., Klingbeil, L., Wieland, M., Eling, C., Schmidtlein, S., Kuhlmann, H., Dikau, R., 2017. Solifluction meets vegetation: the role of biogeomorphic feedbacks for turf-banked solifluction lobe development. Earth Surface Processes and Landforms, 42(11), 1623–1635.

Emmert, A., Kneisel, C., 2015. Internal Structure of Periglacial Landforms: Assessment using 3D Electrical Resistivity Imaging (ERI), EGU General Assembly Conference Abstracts.

Emmert, A., Kneisel, C., 2017. Internal structure of two alpine rock glaciers investigated by quasi-3-D electrical resistivity imaging. The Cryosphere, 11(2), 841–855.

Etzelmüller, B., Farbrot, H., Gudmundsson, Á., Humlum, O., Tveito, O.E., Björnsson, H., 2007. The regional distribution of mountain permafrost in Iceland. Permafrost and Periglacial Processes, 18(2), 185–199.

Etzelmüller, B., Heggem, E.S., Sharkhuu, N., Frauenfelder, R., Kääb, A., Goulden, C., 2006. Mountain permafrost distribution modelling using a multi-criteria approach in the Hövsgöl area, northern Mongolia. Permafrost and Periglacial Processes, 17(2), 91–104.

Everett, M.E., 2013. Ground-penetrating radar. In: M.E. Everett (Ed.), Near-Surface Applied Geophysics. Cambridge University Press, Cambridge, pp. 239–278.

Evin, M., Fabre, D., 1990. The distribution of permafrost in rock glaciers of the Southern Alps (France). Geomorphology, 3(1), 57–71.

Farbrot, H., Etzelmüller, B., Schuler, T.V., Gudmundsson, Á., Eiken, T., Humlum, O., Björnsson, H., 2007. Thermal characteristics and impact of climate change on mountain permafrost in Iceland. Journal of Geophysical Research: Earth Surface, 112(F3).

Farquharson, C.G., Oldenburg, D.W., 2004. A comparison of automatic techniques for estimating the regularization parameter in non-linear inverse problems. Geophysical Journal International, 156(3), 411–425.

Felder, G., 2018. Hier donnert eine Monster-Lawine ins Urserental, Blick.ch, 09.04.2018. Available from https://www.blick.ch/news/schweiz/zentralschweiz/sie-ist-700-meter-breit-und-7-meter-hoch-hier-donnert-eine-monster-lawine-ins-urserental-id8286685.html (last accessed: 22.01.2019).

Feuillet, T., Mercier, D., Decaulne, A., Cossart, E., 2012. Classification of sorted patterned ground areas based on their environmental characteristics (Skagafjördur, Northern Iceland). Geomorphology, 139–140, 577–587.

Flück, W., Hafner, S., Labhart, T.P., 1975. Val Bedretto. Geologischer Atlas der Schweiz 1:25 000. Available from: https://map.geo.admin.ch (last accessed: 22.01.2019).

Fortier, R., LeBlanc, A.-M., Allard, M., Buteau, S., Calmels, F., 2008. Internal structure and conditions of permafrost mounds at Umiujaq in Nunavik, Canada, inferred from field investigation and electrical resistivity tomography. Canadian Journal of Earth Sciences, 45(3), 367–387.

Frauenfelder, R., Kääb, A., 2000. Towards a palaeoclimatic model of rock-glacier formation in the Swiss Alps. Annals of Glaciology, 31, 281–286.

Frehner, M., Ling, A.H.M., Gärtner-Roer, I., 2015. Furrow-and-Ridge Morphology on Rockglaciers Explained by Gravity-Driven Buckle Folding: A Case Study From the Murtèl Rockglacier (Switzerland). Permafrost and Periglacial Processes, 26(1), 57–66.

French, H.M., 2018. The Periglacial Environment. John Wiley & Sons, Hoboken, NJ, 726 pp.

Freppaz, M., Godone, D., Filippa, G., Maggioni, M., Lunardi, S., Williams, M.W., Zanini, E., 2010. Soil erosion caused by snow avalanches: a case study in the Aosta Valley (NW Italy). Arctic, Antarctic, and Alpine Research, 42(4), 412–421.

Fukuda, M., Sone, T., 1992. Some characteristics of alpine permafrost, Mt. Daisetsu, central Hokkaido, northern Japan. Geografiska Annaler. Series A. Physical Geography, 159–167.

Fukui, K., Sone, T., Strelin, J.A., Torielli, C.A., Mori, J., Fujii, Y., 2008. Dynamics and GPR stratigraphy of a polar rock glacier on James Ross Island, Antarctic Peninsula. Journal of Glaciology, 54(186), 445–451.

Funk, M., Hoelzle, M., 1992. A model of potential direct solar radiation for investigating occurrences of mountain permafrost. Permafrost and Periglacial Processes, 3(2), 139–142.

Gensler, G.A., 1978. Das Klima von Graubünden: ein Beitrag zur Regionalklimatologie der Schweiz. Arbeitsberichte der Schweizerischen Meteorologischen Zentralanstalt, 77, 132 pp.

Gharibi, M., Bentley, L.R., 2005. Resolution of 3-D electrical resistivity images from inversions of 2-D orthogonal lines. Journal of Environmental & Engineering Geophysics, 10(4), 339–349.

Glover, P.W., 2016. Archie's law-a reappraisal. Solid Earth, 7(4), 1157–1169.

Godfrey, M.J., Bannister, M.T., Nobes, D., Sletten, R.S., 2008. 3D time-lapse imaging of polygonal patterned ground in the McMurdo dry valleys of Antarctica, 12th International Conference on Ground Penetrating Radar, Birmingham.

Grasmueck, M., Weger, R., 2002. 3D GPR reveals complex internal structure of Pleistocene oolitic sandbar. The Leading Edge, 21(7), 634–639.

Grasmueck, M., Weger, R., Horstmeyer, H., 2004. Three-dimensional ground-penetrating radar imaging of sedimentary structures, fractures, and archaeological features at submeter resolution. Geology, 32(11), 933–936.

Grasmueck, M., Weger, R., Horstmeyer, H., 2005. Full-resolution 3D GPR imaging. Geophysics, 70(1), K12–K19.

Gruber, S., Haeberli, W., 2007. Permafrost in steep bedrock slopes and its temperature-related destabilization following climate change. Journal of Geophysical Research: Earth Surface, 112(F2).

Gubler, S., Fiddes, J., Keller, M., Gruber, S., 2011. Scale-dependent measurement and analysis of ground surface temperature variability in alpine terrain. The Cryosphere, 5(2), 431–443.

Günther, T., Rücker, C., Spitzer, K., 2006. Three-dimensional modelling and inversion of DC resistivity data incorporating topography—II. Inversion. Geophysical Journal International, 166(2), 506–517.

Guo, D., Wang, H., 2016. CMIP5 permafrost degradation projection: A comparison among different regions. Journal of Geophysical Research: Atmospheres, 121(9), 4499–4517.

Gutiérrez, F., Gutiérrez, M., 2016. Desert and Aeolian Landforms. In: F. Gutiérrez, M. Gutiérrez (Eds.), Landforms of the Earth. Springer, pp. 237–259.

Haeberli, W., 1973. Die Basis-Temperatur der winterlichen Schneedecke als möglicher Indikator für die Verbreitung von Permafrost in den Alpen. Zeitschrift für Gletscherkunde und Glazialgeologie, 9, 221–227.

Haeberli, W., 1975. Untersuchungen zur Verbreitung von Permafrost zwischen Flüelapass und Piz Grialetsch (Graubünden). Mitteilungen der Versuchsanstalt für Wasserbau, Hydrologie und Glaziologie an der ETH Zürich, 17, 228 pp.

Haeberli, W., 1985. Creep of mountain permafrost: internal structure and flow of alpine rock glaciers. Mitteilungen der Versuchsanstalt für Wasserbau, Hydrologie und Glaziologie an der ETH Zürich, 77, 142 pp.

Haeberli, W., 1992. Construction, environmental problems and natural hazards in periglacial mountain belts. Permafrost and periglacial processes, 3(2), 111–124.

Haeberli, W., Hallet, B., Arenson, L., Elconin, R., Humlum, O., Kääb, A., Kaufmann, V., Ladanyi, B., Matsuoka, N., Springman, S., 2006. Permafrost creep and rock glacier dynamics. Permafrost and periglacial processes, 17(3), 189–214.

Haeberli, W., Patzelt, G., 1982. Permafrostkartierung im Gebiet der Hochebenkar-Blockgletscher, Obergurgl, Ötztaler Alpen. Zeitschrift für Gletscherkunde und Glazialgeologie, 18(2), 127–150.

Haeberli, W., Vonder Mühll, D., 1996. On the characteristics and possible origins of ice in rock glacier permafrost. Zeitschrift für Geomorphologie. NF Suppl., 104, 43–57.

Hanna, E., Jónsson, T., Box, J.E., 2004. An analysis of Icelandic climate since the nineteenth century. International Journal of Climatology, 24(10), 1193–1210.

Hanson, S., Hoelzle, M., 2004. The thermal regime of the active layer at the Murtèl rock glacier based on data from 2002. Permafrost and Periglacial Processes, 15(3), 273–282.

Harris, C., Arenson, L.U., Christiansen, H.H., Etzelmüller, B., Frauenfelder, R., Gruber, S., Haeberli, W., Hauck, C., Hoelzle, M., Humlum, O., 2009. Permafrost and climate in Europe: Monitoring and modelling thermal, geomorphological and geotechnical responses. Earth-Science Reviews, 92(3), 117–171.

Harris, C., Davies, M.C., Rea, B.R., 2003. Gelifluction: viscous flow or plastic creep? Earth Surface Processes and Landforms: The Journal of the British Geomorphological Research Group, 28(12), 1289–1301.

Harris, C., Kern-Luetschg, M., Smith, F., Isaksen, K., 2008a. Solifluction processes in an area of seasonal ground freezing, Dovrefjell, Norway. Permafrost and Periglacial Processes, 19(1), 31–47.

Harris, C., Smith, J.S., Davies, M.C., Rea, B., 2008b. An investigation of periglacial slope stability in relation to soil properties based on physical modelling in the geotechnical centrifuge. Geomorphology, 93(3–4), 437–459.

Harris, S.A., Pedersen, D.E., 1998. Thermal regimes beneath coarse blocky materials. Permafrost and periglacial processes, 9(2), 107–120.

Hauck, C., 2002. Frozen ground monitoring using DC resistivity tomography. Geophysical Research Letters, 29(21), 12-1–12-4.

Hauck, C., 2013. New concepts in geophysical surveying and data interpretation for permafrost terrain. Permafrost and Periglacial Processes, 24(2), 131–137.

Hauck, C., Bach, M., Hilbich, C., 2008. A 4-phase model to quantify subsurface ice and water content in permafrost regions based on geophysical datasets, Proceedings Ninth International Conference on Permafrost, June, pp. 675–680.

Hauck, C., Vonder Mühll, D., 2003. Inversion and interpretation of two-dimensional geoelectrical measurements for detecting permafrost in mountainous regions. Permafrost and Periglacial Processes, 14(4), 305–318.

Hauck, C., Vonder Mühll, D., Maurer, H., 2003. Using DC resistivity tomography to detect and characterize mountain permafrost. Geophysical Prospecting, 51(4), 273–284.

Haugland, J.E., 2006. Short-term periglacial processes, vegetation succession, and soil development within sorted patterned ground: Jotunheimen, Norway. Arctic, Antarctic, and alpine research, 38(1), 82–89.

Hausmann, H., Krainer, K., Brückl, E., Mostler, W., 2007. Internal structure and ice content of Reichenkar rock glacier (Stubai Alps, Austria) assessed by geophysical investigations. Permafrost and Periglacial Processes, 18(4), 351–367.

Hausmann, H., Krainer, K., Brückl, E., Ullrich, C., 2012. Internal structure, ice content and dynamics of Ölgrube and Kaiserberg rock glaciers (Ötztal Alps, Austria) determined from geophysical surveys. Austrian J. Austrian Journal of Earth Sciences, 105, 12–31.

Hayashi, M., Goeller, N., Quinton, W.L., Wright, N., 2007. A simple heat-conduction method for simulating the frost-table depth in hydrological models. Hydrological Processes, 21 (19), 2610–2622.

Heincke, B., Green, A.G., Kruk, J.v.d., Horstmeyer, H., 2005. Acquisition and processing strategies for 3D georadar surveying a region characterized by rugged topography. Geophysics, 70(6), K53–K61.

Heinrich, G., Gobiet, A., Truhetz, H., Mendlik, T., 2013. Expected climate change and its uncertainty in the Alpine region—extended uncertainty assessment of the reclip: century and ENSEMBLES multi-model dataset. Wegener Center Scientific Reoprt, 50, 69 pp.

Hellman, K., Johansson, S., Olsson, P., Dahlin, T., 2016. Resistivity Inversion Software Comparison, Near Surface Geoscience 2016 — 22nd European Meeting of Environmental and Engineering Geophysics.

Hilbich, C., Fuss, C., Hauck, C., 2011. Automated time-lapse ERT for improved process analysis and monitoring of frozen ground. Permafrost and Periglacial Processes, 22(4), 306–319.

Hilbich, C., Marescot, L., Hauck, C., Loke, M., Mäusbacher, R., 2009. Applicability of electrical resistivity tomography monitoring to coarse blocky and ice-rich permafrost landforms. Permafrost and Periglacial Processes, 20(3), 269–284.

Hiltbrunner, E., 2017. Weather Station ALPFOR. Unpublished Data.

Hinkel, K., Doolittle, J., Bockheim, J., Nelson, F., Paetzold, R., Kimble, J., Travis, R., 2001. Detection of subsurface permafrost features with ground-penetrating radar, Barrow, Alaska. Permafrost and Periglacial Processes, 12(2), 179–190.

Hirakawa, K., 1986. Development of palsa bog in central highland, Iceland. Geographical Reports of Tokyo Metropolitan University (21), 111–122.

Hoelzle, M., 1992. Permafrost occurrence from BTS measurements and climatic parameters in the eastern Swiss Alps. Permafrost and Periglacial Processes, 3(2), 143–147.

Hoelzle, M., Wegmann, M., Krummenacher, B., 1999. Miniature temperature dataloggers for mapping and monitoring of permafrost in high mountain areas: first experience from the Swiss Alps. Permafrost and periglacial processes, 10(2), 113–124.

Hohmann, M., 1997. Soil freezing—the concept of soil water potential. State of the art. Cold Regions Science and Technology, 25(2), 101–110.

Horvath, C.L., 1998. An evaluation of ground penetrating radar for investigation of palsa evolution, MacMillan Pass, NWT, Canada, Proceedings of the Seventh International Conference on Permafrost, Yellowknife, Canada, pp. 473–478.

Hubbard, S.S., Gangodagamage, C., Dafflon, B., Wainwright, H., Peterson, J., Gusmeroli, A., Ulrich, C., Wu, Y., Wilson, C., Rowland, J., 2013. Quantifying and relating land-surface and subsurface variability in permafrost environments using LiDAR and surface geophysical datasets. Hydrogeology Journal, 21(1), 149–169.

Hugenholtz, C.H., Lewkowicz, A.G., 2002. Morphometry and environmental characteristics of turf-banked solifluction lobes, Kluane Range, Yukon Territory, Canada. Permafrost and Periglacial Processes, 13(4), 301–313.

Humlum, O., Christiansen, H.H., Juliussen, H., 2007. Avalanche-derived rock glaciers in Svalbard. Permafrost and Periglacial Processes, 18(1), 75–88.

Icelandic Met Office, 2018. Hveravellir - weather station - information. Available from: https://en.vedur.is/weather/stations/?s=hvell (last accessed: 22.01.2019).

Ikeda, A., 2006. Combination of conventional geophysical methods for sounding the composition of rock glaciers in the Swiss Alps. Permafrost and Periglacial Processes, 17(1), 35–48.

Ikeda, A., Matsuoka, N., 2002. Degradation of talus-derived rock glaciers in the Upper Engadin, Swiss Alps. Permafrost and Periglacial Processes, 13(2), 145–161.

Ikeda, A., Matsuoka, N., 2006. Pebbly versus bouldery rock glaciers: Morphology, structure and processes. Geomorphology, 73(3–4), 279–296.

Ikeda, A., Matsuoka, N., Kääb, A., 2008. Fast deformation of perennially frozen debris in a warm rock glacier in the Swiss Alps: an effect of liquid water. Journal of Geophysical Research: Earth Surface, 113, 12 pp.

Isaksen, K., Ødegård, R.S., Eiken, T., Sollid, J.L., 2000. Composition, flow and development of two tongue-shaped rock glaciers in the permafrost of Svalbard. Permafrost and Periglacial Processes, 11(3), 241–257.

Ishikawa, M., 2003. Thermal regimes at the snow–ground interface and their implications for permafrost investigation. Geomorphology, 52(1–2), 105–120.

Ishikawa, M., Hirakawa, K., 2000. Mountain permafrost distribution based on BTS measurements and DC resistivity soundings in the Daisetsu Mountains, Hokkaido, Japan. Permafrost and Periglacial Processes, 11(2), 109–123.

Iwahana, G., Fukui, K., Mikhailov, N., Ostanin, O., Fujii, Y., 2012. Internal structure of a lithalsa in the Akkol Valley, Russian Altai Mountains. Permafrost and Periglacial Processes, 23(2), 107–118.

Jaesche, P., Veit, H., Huwe, B., 2003. Snow cover and soil moisture controls on solifluction in an area of seasonal frost, eastern Alps. Permafrost and Periglacial Processes, 14(4), 399–410.

Jóhannesson, H., 2014. Geologic Map of Iceland. 1:600 000. Bedrock Geology. Icelandic Institute of Natural History, Reykjavik. Data available from: https://en.isor.is/geological-maps-geological-web-map (last accessed: 22.01.2019).

Jones, D., Harrison, S., Anderson, K., Selley, H., Wood, J., Betts, R., 2018. The distribution and hydrological significance of rock glaciers in the Nepalese Himalaya. Global and Planetary Change, 160, 123–142.

Jones, P.D., Moberg, A., 2003. Hemispheric and large-scale surface air temperature variations: An extensive revision and an update to 2001. Journal of Climate, 16(2), 206–223.

Kääb, A., Frauenfelder, R., Roer, I., 2007. On the response of rockglacier creep to surface temperature increase. Global and Planetary Change, 56(1), 172–187.

Kääb, A., Weber, M., 2004. Development of transverse ridges on rock glaciers: field measurements and laboratory experiments. Permafrost and Periglacial Processes, 15(4), 379–391.

Kane, D.L., Hinkel, K.M., Goering, D.J., Hinzman, L.D., Outcalt, S.I., 2001. Non-conductive heat transfer associated with frozen soils. Global and Planetary Change, 29(3–4), 275–292.

Kasprzak, M., 2015. High-resolution electrical resistivity tomography applied to patterned ground, Wedel Jarlsberg Land, south-west Spitsbergen. Polar Research, 34(1), 25678, 13 pp.

Keller, F., Tamás, M., 2003. Enhanced ground cooling in periods with thin snow cover in the Swiss National Park, Proceedings of the Eighth International Conference on Permafrost, Zurich, Switzerland, pp. 531–536.

Keller, F.U., 1993. Interaktionen zwischen Schnee und Permafrost: eine Grundlagenstudie im Oberengadin, ETH Zürich. 127, Mitteilungen der Versuchsanstalt für Wasserbau, Hydrologie und Glaziologie an der ETH Zürich, 73 pp.

Kenner, R., Magnusson, J., 2017. Estimating the Effect of Different Influencing Factors on Rock Glacier Development in Two Regions in the Swiss Alps. Permafrost and Periglacial Processes, 28(1), 195–208.

Kenner, R., Phillips, M., Hauck, C., Hilbich, C., Mulsow, C., Bühler, Y., Stoffel, A., Buchroithner, M., 2017. New insights on permafrost genesis and conservation in talus slopes based on observations at Flüelapass, Eastern Switzerland. Geomorphology, 290, 101–113.

King, L., Seppälä, M., 1987. Permafrost thickness and distribution in Finnish Lapland — results of geoelectrical soundings. Polarforschung, 57(3), 127–147.

Kinnard, C., Lewkowicz, A.G., 2005. Movement, moisture and thermal conditions at a turf-banked solifluction lobe, Kluane Range, Yukon Territory, Canada. Permafrost and Periglacial Processes, 16(3), 261–275.

Kneisel, C., 1999. Permafrost in Gletschervorfeldern eine vergleichende Untersuchung in den Ostschweizer Alpen und Nordschweden. Trierer Geographische Studien, 22, 158 pp.

Kneisel, C., 2006. Assessment of subsurface lithology in mountain environments using 2D resistivity imaging. Geomorphology, 80(1–2), 32–44.

Kneisel, C., 2010. The nature and dynamics of frozen ground in alpine and subarctic periglacial environments. The Holocene, 20(3), 423–445.

Kneisel, C., Emmert, A., Kästl, J., 2014a. Application of 3D electrical resistivity imaging for mapping frozen ground conditions exemplified by three case studies. Geomorphology, 210, 71–82.

Kneisel, C., Emmert, A., Polich, P., Zollinger, B., Egli, M., 2015. Soil geomorphology and frozen ground conditions at a subalpine talus slope having permafrost in the eastern Swiss Alps. Catena, 133, 107–118.

Kneisel, C., Hauck, C., 2008. Electrical methods. In: C. Kneisel, C. Hauck (Eds.), Applied Geophysics in Periglacial Environments. Cambridge University Press, Cambridge, pp. 3–27.

Kneisel, C., Hauck, C., Fortier, R., Moorman, B., 2008. Advances in geophysical methods for permafrost investigations. Permafrost and Periglacial Processes, 19(2), 157–178.

Kneisel, C., Hauck, C., Mühll, D.V., 2000. Permafrost below the timberline confirmed and characterized by geoelectrical resistivity measurements, Bever Valley, eastern Swiss Alps. Permafrost and Periglacial Processes, 11(4), 295–304.

Kneisel, C., Kääb, A., 2007. Mountain permafrost dynamics within a recently exposed glacier forefield inferred by a combined geomorphological, geophysical and photogrammetrical approach. Earth Surface Processes and Landforms: The Journal of the British Geomorphological Research Group, 32(12), 1797–1810.

Kneisel, C., Rödder, T., Schwindt, D., 2014b. Frozen ground dynamics resolved by multiyear and yearround electrical resistivity monitoring at three alpine sites in the Swiss Alps. Near Surface Geophysics, 12(1), 117–132.

Kneisel, C., Saemundsson, T., Beylich, A., 2007. Reconnaissance surveys of contemporary permafrost environments in central Iceland using geoelectrical methods: implications for permafrost degradation and sediment fluxes. Geografiska Annaler: Series A, Physical Geography, 89(1), 41–50.

Knight, J., Harrison, S., Jones, D.B., 2018. Rock glaciers and the geomorphological evolution of deglacierizing mountains. Geomorphology, 324, 14–24.

Kohout, T., Bućko, M.S., Rasmus, K., Leppäranta, M., Matero, I., 2014. Non-Invasive Geophysical Investigation and Thermodynamic Analysis of a Palsa in Lapland, Northwest Finland. Permafrost and Periglacial Processes, 25(1), 45–52.

Kovats, R.S., Valentini, R., Bouwer, L.M., Georgopoulou, E., Jacob, D., Martin, E., Rounsevell, M., Soussana, J.F., 2014. Europe. In: V.R. Barros, C.B. Field, D.J. Dokken, M.D. Mastrandrea, K.J. Mach, T.E. Bilir, M. Chatterjee, K.L. Ebi, Y.O. Estrada, R.C. Genova, B. Girma, E.S. Kissel, A.N. Levy, S. MacCracken, P.R. Mastrandrea, L.L. White (Eds.), Climate Change 2014: Impacts, Adaptation, and Vulnerability. Part B: Regional Aspects. Contribution of Working Group II to the Fifth Assessment Report of the Intergovernmental Panel of Climate Change. Cambridge University Press, Cambridge, United Kingdom and New York, NY, USA, pp. 1267–1326.

Krainer, K., Mostler, W., 2002. Hydrology of active rock glaciers: examples from the Austrian Alps. Arctic, Antarctic, and Alpine Research, 34(2), 142–149.

Krainer, K., Mussner, L., Behm, M., Hausmann, H., 2012. Multi-disciplinary investigation of an active rock glacier in the Sella group (Dolomites; Northern Italy). Austrian Journal of Earth Sciences, 105(2).

Krautblatter, M., Verleysdonk, S., Flores-Orozco, A., Kemna, A., 2010. Temperature-calibrated imaging of seasonal changes in permafrost rock walls by quantitative electrical resistivity tomography (Zugspitze, German/Austrian Alps). Journal of Geophysical Research: Earth Surface, 115(F2), 48–62.

Kristensen, L., Juliussen, H., Christiansen, H.H., Humlum, O., 2009. Structure and composition of a tidewater glacier push moraine, Svalbard, revealed by DC resistivity profiling. Boreas, 38(1), 176–186.

Kuhry, P., 2008. Palsa and peat plateau development in the Hudson Bay Lowlands, Canada: timing, pathways and causes. Boreas, 37(2), 316–327.

Kujala, K., Seppälä, M., Holappa, T., 2008. Physical properties of peat and palsa formation. Cold Regions Science and Technology, 52(3), 408–414.

Labhart, T.P., Isler, A., Renner, F.B., 2012. Urseren. Geologischer Atlas der Schweiz 1:25 000. Available from: https://map.geo.admin.ch/ (last accessed: 22.01.2019).

Lambiel, C., Delaloye, R., 2004. Contribution of real-time kinematic GPS in the study of creeping mountain permafrost: Examples from the Western Swiss Alps. Permafrost and periglacial processes, 15(3), 229–241.

Lambiel, C., Pieracci, K., 2008. Permafrost distribution in talus slopes located within the alpine periglacial belt, Swiss Alps. Permafrost and Periglacial Processes, 19(3), 293–304.

Lambot, S., André, F., Slob, E., Vereecken, H., 2012. Effect of antenna-medium coupling in the analysis of ground-penetrating radar data. Near Surface Geophysics, 10(6), 631–639.

Langston, G., Bentley, L.R., Hayashi, M., McClymont, A., Pidlisecky, A., 2011. Internal structure and hydrological functions of an alpine proglacial moraine. Hydrological Processes, 25(19), 2967–2982.

Larsen, J.N., Anisimov, O.A., Constable, A., Hollowed, A.B., Maynard, N., Prestrud, P., Prowse, T.D., Stone, J.M.R., 2014. Polar regions. In: V.R. Barros, C.B. Field, D.J. Dokken, M.D. Mastrandrea, K.J. Mach, T.E. Bilir, M. Chatterjee, K.L. Ebi, Y.O. Estrada, R.C. Genova, B. Girma, E.S. Kissel, A.N. Levy, S. MacCracken, P.R. Mastrandrea, L.L. White (Eds.), Climate Change 2014: Impacts, Adaptation, and Vulnerability. Part B: Regional Aspects. Contribution of Working Group II to the Fifth Assessment Report of the Intergovernmental Panel of Climate Change. Cambridge University Press, Cambridge, United Kingdom and New York, NY, USA, pp. 1567–1612.

Laternser, M., Schneebeli, M., 2003. Long-term snow climate trends of the Swiss Alps (1931–99). International Journal of Climatology, 23(7), 733–750.

Lawrence, D.M., Slater, A.G., 2005. A projection of severe near-surface permafrost degradation during the 21st century. Geophysical Research Letters, 32(24), 5 pp.

Lawrence, D.M., Slater, A.G., Swenson, S.C., 2012. Simulation of present-day and future permafrost and seasonally frozen ground conditions in CCSM4. Journal of Climate, 25(7), 2207-2225.

Léger, E., Dafflon, B., Soom, F., Peterson, J., Ulrich, C., Hubbard, S., 2017. Quantification of Arctic Soil and Permafrost Properties Using Ground-Penetrating Radar and Electrical Resistivity Tomography Datasets. IEEE Journal of Selected Topics in Applied Earth Observations and Remote Sensing, 10(10), 4348–4359.

Leopold, M., Dethier, D., Völkel, J., Raab, T., Rikert, T.C., Caine, N., 2008. Using geophysical methods to study the shallow subsurface of a sensitive alpine environment, Niwot Ridge, Colorado Front Range, USA. Arctic, Antarctic, and Alpine Research, 40(3), 519–530.

Leopold, M., Williams, M., Caine, N., Völkel, J., Dethier, D., 2011. Internal structure of the Green lake 5 rock glacier, Colorado Front Range, USA. Permafrost and Periglacial Processes, 22(2), 107–119.

Lerjen, M., Kääb, A., Hoelzle, M., Haeberli, W., 2003. Local distribution pattern of discontinuous mountain permafrost. A process study at Flüela Pass, Swiss Alps, Proceedings of the 8th International Conference on Permafrost, pp. 21–25.

Lewkowicz, A.G., 2008. Evaluation of miniature temperature-loggers to monitor snowpack evolution at mountain permafrost sites, northwestern Canada. Permafrost and Periglacial Processes, 19(3), 323–331.

Lewkowicz, A.G., Etzelmüller, B., Smith, S.L., 2011. Characteristics of discontinuous permafrost based on ground temperature measurements and electrical resistivity tomography, southern Yukon, Canada. Permafrost and Periglacial Processes, 22(4), 320–342.

Li, Y., Oldenburg, D.W., 1999. 3-D inversion of DC resistivity data using an L-curve criterion, SEG Technical Program Expanded Abstracts 1999. Society of Exploration Geophysicists, pp. 251–254.

Loke, M., 2016a. Rapid 3-D Resistivity & IP Inversion using the least-squares method (For 3-D surveys using the pole-pole, pole-dipole, dipole-dipole, rectangular, Wenner, Wenner-Schlumberger and non-conventional arrays) On land, aquatic, cross-borehole and time-lapse surveys. Geotomo Software. Manual, March 2016.

Loke, M., 2016b. Tutorial: 2-D and 3-D electrical imaging surveys (Rev. date: 26th June 2016). Available from: http://geotomosoft.com/coursenotes.zip (last accessed: 10.01.2017).

Loke, M., Chambers, J., Rucker, D., Kuras, O., Wilkinson, P., 2013. Recent developments in the direct-current geoelectrical imaging method. Journal of Applied Geophysics, 95, 135–156.

Loke, M.H., Acworth, I., Dahlin, T., 2003. A comparison of smooth and blocky inversion methods in 2D electrical imaging surveys. Exploration Geophysics, 34(3), 182–187.

Lowrie, W., 2007. Fundamentals of geophysics. Cambridge University Press, Cambridge, 381 pp.

Lowrie, W., 2011. A student's guide to geophysical equations. Cambridge University Press, Cambridge, 296 pp.

Luetschg, M., Stoeckli, V., Lehning, M., Haeberli, W., Ammann, W., 2004. Temperatures in two boreholes at Flüela Pass, Eastern Swiss Alps: the effect of snow redistribution on permafrost distribution patterns in high mountain areas. Permafrost and Periglacial Processes, 15(3), 283–297.

Lugon, R., Delaloye, R., Serrano, E., Reynard, E., Lambiel, C., González-Trueba, J.J., 2004. Permafrost and little ice age glacier relationships, Posets Massif, Central Pyrenees, Spain. Permafrost and Periglacial Processes, 15(3), 207–220.

Luoto, M., Hjort, J., 2006. Scale matters–A multi-resolution study of the determinants of patterned ground activity in subarctic Finland. Geomorphology, 80(3–4), 282–294.

Marescot, L., Loke, M., Chapellier, D., Delaloye, R., Lambiel, C., Reynard, E., 2003. Assessing reliability of 2D resistivity imaging in mountain permafrost studies using the depth of investigation index method. Near Surface Geophysics, 1(2), 57–67.

Marklund, P., 2014. A 2D Electrical Resistivity Survey of Palsas in Tavvavuoma, sub-arctic Sweden. M.Sc. Master Thesis, University of Uppsala, Uppsala, 72 pp.

Marmy, A., Salzmann, N., Scherler, M., Hauck, C., 2013. Permafrost model sensitivity to seasonal climatic changes and extreme events in mountainous regions. Environmental Research Letters, 8(3), 035048, 9 pp.

Matsumoto, H., Ishikawa, M., 2002. Gelifluction within a solifluction lobe in the Kärkevagge valley, Swedish Lapland. Geografiska Annaler: Series A, Physical Geography, 84(3–4), 261–266.

Matsuoka, N., 1996. Soil moisture variability in relation to diurnal frost heaving on Japanese high mountain slopes. Permafrost and Periglacial Processes, 7(2), 139–151.

Matsuoka, N., 2001. Solifluction rates, processes and landforms: a global review. Earth-Science Reviews, 55(1–2), 107–134.

Matsuoka, N., 2008. Frost weathering and rockwall erosion in the southeastern Swiss Alps: Long-term (1994–2006) observations. Geomorphology, 99(1–4), 353–368.

Matsuoka, N., 2010. Solifluction and mudflow on a limestone periglacial slope in the Swiss Alps: 14 years of monitoring. Permafrost and Periglacial Processes, 21(3), 219–240.

Matsuoka, N., 2014. Combining time-lapse photography and multisensor monitoring to understand frost creep dynamics in the Japanese Alps. Permafrost and Periglacial Processes, 25(2), 94–106.

Matsuoka, N., Abe, M., Ijiri, M., 2003. Differential frost heave and sorted patterned ground: field measurements and a laboratory experiment. Geomorphology, 52(1–2), 73–85.

Matsuoka, N., Ikeda, A., Date, T., 2005. Morphometric analysis of solifluction lobes and rock glaciers in the Swiss Alps. Permafrost and Periglacial Processes, 16(1), 99–113.

Matthews, J.A., Dahl, S.O., Berrisford, M.S., Nesje, A., 1997. Cyclic development and thermokarstic degradation of palsas in the mid-alpine zone at Leirpullan, Dovrefjell, southern Norway. Permafrost and periglacial processes, 8(1), 107–122.

Matthews, J.A., Shakesby, R.A., Berrisford, M.S., McEwen, L.J., 1998. Periglacial patterned ground on the Styggedalsbreen glacier foreland, Jotunheimen, southern Norway: micro-topographic, paraglacial and geoecological controls. Permafrost and Periglacial Processes, 9(2), 147–166.

Maurer, H., Hauck, C., 2007. Geophysical imaging of alpine rock glaciers. Journal of Glaciology, 53(180), 110–120.

Mellon, M.T., Arvidson, R.E., Marlow, J.J., Phillips, R.J., Asphaug, E., 2008. Periglacial landforms at the Phoenix landing site and the northern plains of Mars. Journal of Geophysical Research: Planets (1991–2012), 113(E3), 15 pp.

Merz, K., Maurer, H., Rabenstein, L., Buchli, T., Springman, S.M., Zweifel, M., 2015. Multidisciplinary geophysical investigations over an alpine rock glacier. Geophysics, 81(1), WA147–WA157.

MeteoSchweiz, 2018. Database IDAweb. Bundesamt für Meteorologie und Klimatologie Zürich. Available from: https://gate.meteoswiss.ch/idaweb (last accessed 07.12.2018).

Miyamoto, H., Haruyama, J.i., Kobayashi, T., Suzuki, K., Okada, T., Nishibori, T., Showman, A.P., Lorenz, R., Mogi, K., Crown, D.A., 2005. Mapping the structure and depth of lava tubes using ground penetrating radar. Geophysical research letters, 32(21), 5 pp.

Monnier, S., Camerlynck, C., Rejiba, F., 2008. Ground penetrating radar survey and stratigraphic interpretation of the Plan du Lac rock glaciers, Vanoise Massif, northern French Alps. Permafrost and Periglacial Processes, 19(1), 19–30.

Monnier, S., Camerlynck, C., Rejiba, F., Kinnard, C., Feuillet, T., Dhemaied, A., 2011. Structure and genesis of the Thabor rock glacier (Northern French Alps) determined from morphological and ground-penetrating radar surveys. Geomorphology, 134(3–4), 269–279.

Monnier, S., Camerlynck, C., Rejiba, F., Kinnard, C., Galibert, P.Y., 2013. Evidencing a large body of ice in a rock glacier, Vanoise Massif, Northern French Alps. Geografiska Annaler: Series A, Physical Geography, 95(2), 109–123.

Monnier, S., Kinnard, C., 2015. Internal Structure and Composition of a Rock Glacier in the Dry Andes, Inferred from Ground-penetrating Radar Data and its Artefacts. Permafrost and Periglacial Processes, 26(4), 335–346.

Monnier, S., Kinnard, C., Surazakov, A., Bossy, W., 2014. Geomorphology, internal structure, and successive development of a glacier foreland in the semiarid Chilean Andes (Cerro Tapado, upper Elqui Valley, 30 08' S., 69 55' W.). Geomorphology, 207, 126–140.

Moorman, B.J., Michel, F.A., 2000. The burial of ice in the proglacial environment on Bylot Island, Arctic Canada. Permafrost and Periglacial Processes, 11(3), 161–175.

Moorman, B.J., Robinson, S.D., Burgess, M.M., 2003. Imaging periglacial conditions with ground-penetrating radar. Permafrost and Periglacial Processes, 14(4), 319–329.

Mori, J., Fukui, K., Sone, T., Strelin, J., Torielli, C., 2007. Internal structure of stone-banked lobes and terraces on Rink Plateau, James Ross Island, Antarctic Peninsula region. Polish Polar Research, 28(1), 23–30.

Müller, J., Vieli, A., Gärtner-Roer, I., 2016. Rock glaciers on the run – understanding rock glacier landform evolution and recent changes from numerical flow modeling. The Cryosphere, 10(6), 2865–2886.

Munroe, J.S., Doolittle, J.A., Kanevskiy, M.Z., Hinkel, K.M., Nelson, F.E., Jones, B.M., Shur, Y., Kimble, J.M., 2007. Application of ground-penetrating radar imagery for three-dimensional visualisation of near-surface structures in ice-rich permafrost, Barrow, Alaska. Permafrost and Periglacial Processes, 18(4), 309–321.

Nakicenovic, N., Alcamo, J., Grubler, A., Riahi, K., Roehrl, R., Rogner, H.-H., Victor, N., 2000. Special report on emissions scenarios (SRES), a special report of Working Group III of the intergovernmental panel on climate change. Cambridge University Press, Cambridge, 608 pp.

Neal, A., 2004. Ground-penetrating radar and its use in sedimentology: principles, problems and progress. Earth-science reviews, 66(3–4), 261–330.

Oddur, S., Williams, R.S., Vikingsson, S., 2013. Jöklakort af Islandi 1:500 000. Utgefid af Vedurstofu Islands.

Oksanen, P.O., 2006. Holocene development of the Vaisjeäggi palsa mire, Finnish Lapland. Boreas, 35(1), 81–95.

Osterkamp, T., Jurick, R., Gislason, G., Akasofu, S.-I., 1980. Electrical resistivity measurements in permafrost terrain at the Engineer Creek road cut, Fairbanks, Alaska. Cold Regions Science and Technology, 3(4), 277–286.

Østrem, G., 1964. Ice-cored moraines in Scandinavia. Geografiska Annaler, 46(3), 282–337.

Otto, J., Sass, O., 2006. Comparing geophysical methods for talus slope investigations in the Turtmann valley (Swiss Alps). Geomorphology, 76(3–4), 257–272.

Outcalt, S.I., Benedict, J.B., 1965. Photo-interpretation of two types of rock glacier in the Colorado Front Range, USA. Journal of Glaciology, 5(42), 849–856.

Peel, M.C., Finlayson, B.L., McMahon, T.A., 2007. Updated world map of the Köppen-Geiger climate classification. Hydrol. Earth Syst. Sci., 11(5), 1633–1644.

Pellet, C., Hilbich, C., Marmy, A., Hauck, C., 2016. Soil Moisture Data for the Validation of Permafrost Models Using Direct and Indirect Measurement Approaches at Three Alpine Sites. Frontiers in Earth Science, 3(91).

Peters, T., Gouffon, Y., Cornelius, H.P., Staub, R., 2005. St. Moritz/ S. Murezzan. Geologischer Atlas der Schweiz 1:25 000. Available from: https://map.geo.admin.ch/ (last accessed: 22.01.2019).

Peterson, R.A., Krantz, W.B., 2008. Differential frost heave model for patterned ground formation: Corroboration with observations along a North American arctic transect. Journal of Geophysical Research: Biogeosciences (2005–2012), 113(G3), 17 pp.

Pfiffner, O.A., 2015. Geologie der Alpen. utb 8416, Haupt, Bern, 400 pp.

Pidlisecky, A., Haber, E., Knight, R., 2007. RESINVM3D: A 3D resistivity inversion package. Geophysics, 72(2), H1–H10.

Pissart, A., 2002. Palsas, lithalsas and remnants of these periglacial mounds. A progress report. Progress in Physical Geography, 26(4), 605–621.

Pruessner, L., Phillips, M., Farinotti, D., Hoelzle, M., Lehning, M., 2018. Near-surface ventilation as a key for modeling the thermal regime of coarse blocky rock glaciers. Permafrost and Periglacial Processes, 29(3), 152–163.

Räisänen, J., Hansson, U., Ullerstig, A., Döscher, R., Graham, L., Jones, C., Meier, H., Samuelsson, P., Willén, U., 2004. European climate in the late twenty-first century: regional simulations with two driving global models and two forcing scenarios. Climate dynamics, 22(1), 13–31.

Reusser, D.E., Zehe, E., 2011. Low-cost monitoring of snow height and thermal properties with inexpensive temperature sensors. Hydrological Processes, 25(12), 1841–1852.

Reynolds, J., Paren, J., 1980. Recrystallisation and electrical behaviour of glacier ice. Nature, 283(5742), 63.

Reynolds, J.M., 2011. An introduction to applied and environmental geophysics. John Wiley & Sons, Hoboken, NJ, 710 pp.

Rhoads, B.L., Thorn, C.E., 1996. The scientific nature of geomorphology, Proceedings of the 27th Binghamton Symposium in Geomorphology, pp. 27–29.

Riahi, K., Rao, S., Krey, V., Cho, C., Chirkov, V., Fischer, G., Kindermann, G., Nakicenovic, N., Rafaj, P., 2011. RCP 8.5—A scenario of comparatively high greenhouse gas emissions. Climatic Change, 109(1), 33–57.

Ribolini, A., Chelli, A., Guglielmin, M., Pappalardo, M., 2007. Relationships between glacier and rock glacier in the Maritime Alps, Schiantala Valley, Italy. Quaternary Research, 68(3), 353–363.

Ribolini, A., Guglielmin, M., Fabre, D., Bodin, X., Marchisio, M., Sartini, S., Spagnolo, M., Schoeneich, P., 2010. The internal structure of rock glaciers and recently deglaciated slopes as revealed by geoelectrical tomography: insights on permafrost and recent glacial evolution in the Central and Western Alps (Italy–France). Quaternary Science Reviews, 29(3–4), 507–521.

Ridefelt, H., Boelhouwers, J., 2006. Observations on regional variation in solifluction landform morphology and environment in the Abisko region, northern Sweden. Permafrost and Periglacial Processes, 17(3), 253–266.

Rist, A., Phillips, M., 2005. First results of investigations on hydrothermal processes within the active layer above alpine permafrost in steep terrain. Norsk Geografisk Tidsskrift-Norwegian Journal of Geography, 59(2), 177–183.

Rödder, T., Kneisel, C., 2012a. Influence of snow cover and grain size on the ground thermal regime in the discontinuous permafrost zone, Swiss Alps. Geomorphology, 175, 176–189.

Rödder, T., Kneisel, C., 2012b. Permafrost mapping using quasi-3D resistivity imaging, Murtèl, Swiss Alps. Near Surface Geophysics, 10(2), 117–127.

Roer, I., Kääb, A., Dikau, R., 2005. Rockglacier acceleration in the Turtmann valley (Swiss Alps): Probable controls. Norsk Geografisk Tidsskrift-Norwegian Journal of Geography, 59(2), 157–163.

Ross, N., Brabham, P.J., Harris, C., Christiansen, H.H., 2007. Internal structure of open system pingos, Adventdalen, Svalbard: the use of resistivity tomography to assess ground-ice conditions. Journal of Environmental and Engineering Geophysics, 12(1), 113–126.

Rücker, C., Günther, T., Spitzer, K., 2006. Three-dimensional modelling and inversion of DC resistivity data incorporating topography—I. Modelling. Geophysical Journal International, 166(2), 495–505.

Saemundsson, T., Arnalds, O., Kneisel, C., Jonsson, H.P., Decaulne, A., 2012. The Orravatnsrustir palsa site in Central Iceland—Palsas in an aeolian sedimentation environment. Geomorphology, 167, 13–20.

Sandmeier, K.J., 2017. REFLEXW Version 8.5 manual.

Sasaki, Y., 1992. Resolution of Resistivity Tomography Inferred from Numerical Simulation 1. Geophysical prospecting, 40(4), 453–463.

Sass, O., 2006. Determination of the internal structure of alpine talus deposits using different geophysical methods (Lechtaler Alps, Austria). Geomorphology, 80(1–2), 45–58.

Sass, O., 2007. Bedrock detection and talus thickness assessment in the European Alps using geophysical methods. Journal of Applied Geophysics, 62(3), 254–269.

Sawada, Y., Ishikawa, M., Ono, Y., 2003. Thermal regime of sporadic permafrost in a block slope on Mt. Nishi-Nupukaushinupuri, Hokkaido Island, Northern Japan. Geomorphology, 52(1–2), 121–130.

Scapozza, C., Laigre, L., 2014. The contribution of Electrical Resistivity Tomography (ERT) in Alpine dynamics geomorphology: case studies from the Swiss Alps. Géomorphologie: relief, processus, environnement, 20(1), 27–42.

Scapozza, C., Lambiel, C., Baron, L., Marescot, L., Reynard, E., 2011. Internal structure and permafrost distribution in two alpine periglacial talus slopes, Valais, Swiss Alps. Geomorphology, 132(3–4), 208–221.

Schennen, S., Tronicke, J., Wetterich, S., Allroggen, N., Schwamborn, G., Schirrmeister, L., 2016. 3D ground-penetrating radar imaging of ice complex deposits in northern East Siberia. Geophysics, 81(1), WA195–WA202.

Scherler, M., Hauck, C., Hoelzle, M., Stähli, M., Völksch, I., 2010. Meltwater infiltration into the frozen active layer at an alpine permafrost site. Permafrost and Periglacial Processes, 21(4), 325–334.

Scherler, M., Schneider, S., Hoelzle, M., Hauck, C., 2014. A two-sided approach to estimate heat transfer processes within the active layer of the Murtèl–Corvatsch rock glacier. Earth Surface Dynamics, 2(1), 141–154.

Schmid, M., Baral, P., Gruber, S., Shahi, S., Shrestha, T., Stumm, D., Wester, P., 2015. Assessment of permafrost distribution maps in the Hindu Kush Himalayan region using rock glaciers mapped in Google Earth. Cryosphere, 9(6), 2089–2099.

Schneider, S., Daengeli, S., Hauck, C., Hoelzle, M., 2013. A spatial and temporal analysis of different periglacial materials by using geoelectrical, seismic and borehole temperature data at Murtèl–Corvatsch, Upper Engadin, Swiss Alps. Geographica Helvetica, 68(4), 265–280.

Schneider, S., Hoelzle, M., Hauck, C., 2012. Influence of surface and subsurface heterogeneity on observed borehole temperatures at a mountain permafrost site in the Upper Engadine, Swiss Alps. The Cryosphere, 6(2), 517–531.

Schrott, L., Hufschmidt, G., Hankammer, M., Hoffmann, T., Dikau, R., 2003. Spatial distribution of sediment storage types and quantification of valley fill deposits in an alpine basin, Reintal, Bavarian Alps, Germany. Geomorphology, 55(1–4), 45–63.

Schrott, L., Sass, O., 2008. Application of field geophysics in geomorphology: advances and limitations exemplified by case studies. Geomorphology, 93(1–2), 55–73.

Schwamborn, G., Heinzel, J., Schirrmeister, L., 2008. Internal characteristics of ice-marginal sediments deduced from georadar profiling and sediment properties (Brøgger Peninsula, Svalbard). Geomorphology, 95(1–2), 74–83.

Scott, W., Sellmann, P., Hunter, J., 1990. Geophysics in the study of permafrost. Geotechnical and environmental geophysics, 1, 355–384.

Seppälä, M., 1972. The term 'palsa'. Zeitschrift für Geomorphologie, 16(4), 463.

Seppälä, M., 1986. The origin of palsas. Geografiska Annaler: Series A, Physical Geography, 68(3), 141–147.

Seppälä, M., 2003. Surface abrasion of palsas by wind action in Finnish Lapland. Geomorphology, 52(1–2), 141–148.

Seppälä, M., 2011. Synthesis of studies of palsa formation underlining the importance of local environmental and physical characteristics. Quaternary Research, 75(2), 366–370.

Seppälä, M., Kujala, K., 2009. The role of buoyancy in palsa formation. Geological Society, London, Special Publications, 320(1), 51–56.

Seppi, R., Zanoner, T., Carton, A., Bondesan, A., Francese, R., Carturan, L., Zumiani, M., Giorgi, M., Ninfo, A., 2015. Current transition from glacial to periglacial processes in the Dolomites (South-Eastern Alps). Geomorphology, 228, 71–86.

Serquet, G., Marty, C., Dulex, J.-P., Rebetez, M., 2011. Seasonal trends and temperature dependence of the snowfall/precipitation-day ratio in Switzerland. Geophysical Research Letters, 38(7).

Shilts, W.W., 1978. Nature and genesis of mudboils, central Keewatin, Canada. Canadian Journal of Earth Sciences, 15(7), 1053–1068.

Shur, Y., Hinkel, K.M., Nelson, F.E., 2005. The transient layer: implications for geocryology and climate-change science. Permafrost and Periglacial Processes, 16(1), 5–17.

Sjöberg, Y., Coon, E., Sannel, A.B.K., Pannetier, R., Harp, D., Frampton, A., Painter, S.L., Lyon, S.W., 2016. Thermal effects of groundwater flow through subarctic fens: A case study based on field observations and numerical modeling. Water Resources Research, 52(3), 1591–1606.

Sjöberg, Y., Marklund, P., Pettersson, R., Lyon, S.W., 2015. Geophysical mapping of palsa peatland permafrost. The Cryosphere, 9(2), 465–478.

Sjödahl, P., Dahlin, T., Zhou, B., 2006. 2.5 D resistivity modeling of embankment dams to assess influence from geometry and material properties. Geophysics, 71(3), G107–G114.

Smith, D.J., 1988. Rates and controls of soil movement on a solifluction slope in the Mount Rae area, Canadian Rocky Mountains. Zeitschrift für Geomorphologie NF, 71, 25–44.

Sovilla, B., McElwaine, J.N., Schaer, M., Vallet, J., 2010. Variation of deposition depth with slope angle in snow avalanches: Measurements from Vallée de la Sionne. Journal of Geophysical Research: Earth Surface, 115(F2), 13 pp.

Springman, S.M., Arenson, L.U., Yamamoto, Y., Maurer, H., Kos, A., Buchli, T., Derungs, G., 2012. Multidisciplinary investigations on three rock glaciers in the Swiss Alps: legacies and future perspectives. Geografiska Annaler: Series A, Physical Geography, 94(2), 215–243.

Squyres, S.W., Carr, M.H., 1986. Geomorphic evidence for the distribution of ground ice on Mars. Science, 231(4735), 249–252.

Stahr, A.R., Langenscheidt, E., 2015. Landforms of high mountains. Springer, Heidelberg, 152 pp.

Steedman, A.E., Lantz, T.C., Kokelj, S.V., 2017. Spatio-Temporal Variation in High-Centre Polygons and Ice-Wedge Melt Ponds, Tuktoyaktuk Coastlands, Northwest Territories. Permafrost and Periglacial Processes, 28, 66–78.

Stenni, B., Genoni, L., Flora, O., Guglielmin, M., 2007. An oxygen isotope record from the Foscagno rock-glacier ice core, Upper Valtellina, Italian Central Alps. The Holocene, 17(7), 1033–1039.

Stevens, C.W., Moorman, B.J., Solomon, S.M., 2008. Detection of frozen and unfrozen interfaces with ground penetrating radar in the nearshore zone of the Mackenzie Delta, Canada, Proceedings of the 9th International Conference on Permafrost, pp. 1711–1716.

Stilla U., Hoegner L., 2007. Automatische Auswertung von Wärmebildsequenzen. DVW Bayern Mitteilungen 59(1), 51–62.

Stummer, P., Maurer, H., Green, A.G., 2004. Experimental design: Electrical resistivity data sets that provide optimum subsurface information. Geophysics, 69(1), 120–139.

Stummer, P., Maurer, H., Horstmeyer, H., Green, A.G., 2002. Optimization of DC resistivity data acquisition: Real-time experimental design and a new multielectrode system. IEEE Transactions on Geoscience and Remote Sensing, 40(12), 2727–2735.

swisstopo, 2005. DHM25. Das digitale Höhenmodell der Schweiz. Bundesamt für Landestopografie, Wabern, Digital Data Set.

Telford, W.M., Telford, W., Geldart, L., Sheriff, R.E., Sheriff, R., 1990. Applied geophysics. Cambridge University Press, 790 pp.

Tran, A.P., Dafflon, B., Bisht, G., Hubbard, S.S., 2018. Spatial and temporal variations of thaw layer thickness and its controlling factors identified using time-lapse electrical resistivity tomography and hydro-thermal modeling. Journal of Hydrology, 561, 751–763.

Van Dam, R.L., 2012. Landform characterization using geophysics—Recent advances, applications, and emerging tools. Geomorphology, 137(1), 57–73.

Van der Kruk, J., 2015. 11.07 - Tools and Techniques: Ground-Penetrating Radar. In: G. Schubert (Ed.), Treatise on Geophysics (Second Edition). Elsevier, Oxford, pp. 209–232.

Van Everdingen, R., 2005. Multi-language glossary of permafrost and related ground-ice terms in Chinese, English, French, German, Icelandic, Italian, Norwegian, Polish, Romanian, Russian, Spanish, and Swedish. International Permafrost Association. Available from: https://globalcryospherewatch.org/reference/glossary_docs/Glossary_of_Permafrost_and_Ground-Ice_IPA_2005.pdf (last accessed: 22.01.2019).

Van Overmeeren, R., 1998. Radar facies of unconsolidated sediments in The Netherlands: A radar stratigraphy interpretation method for hydrogeology. Journal of Applied Geophysics, 40(1–3), 1–18.

Van Vliet-Lanoë, B., 1988. The significance of cryoturbation phenomena in environmental reconstruction. Journal of Quaternary Science, 3(1), 85–96.

Vanhala, H., Lintinen, P., Ojala, A., 2009. Electrical resistivity study of permafrost on Ridnitšohkka fell in northwest Lapland, Finland. Geophysica, 45(1-2), 103–118.

Villarroel, C., Tamburini Beliveau, G., Forte, A., Monserrat, O., Morvillo, M., 2018. DInSAR for a Regional Inventory of Active Rock Glaciers in the Dry Andes Mountains of Argentina and Chile with Sentinel-1 Data. Remote Sensing, 10(10), 1588, 21 pp.

Vonder Mühll, D., Hauck, C., Gubler, H., 2002. Mapping of mountain permafrost using geophysical methods. Progress in Physical Geography, 26(4), 643–660.

Vonder Mühll, D., Hauck, C., Lehmann, F., 2000. Verification of geophysical models in Alpine permafrost using borehole information. Annals of Glaciology, 31, 300–306.

Vonder Mühll, D.S., 1992. Evidence of intrapermafrost groundwater flow beneath an active rock glacier in the Swiss Alps. Permafrost and Periglacial Processes, 3(2), 169–173.

Wahrhaftig, C., Cox, A., 1959. Rock glaciers in the Alaska Range. Geological Society of America Bulletin, 70(4), 383–436.

Wakonigg, H., 1996. Unterkühlte Schutthalden, Arbeiten aus dem Insitut für Geographie der Karl-Franzens Universität, Graz, pp. 209–223.

Walker, D.A., Epstein, H.E., Gould, W.A., Kelley, A.M., Kade, A.N., Knudson, J.A., Krantz, W.B., Michaelson, G., Peterson, R.A., Ping, C.L., 2004. Frost-boil ecosystems: complex interactions between landforms, soils, vegetation and climate. Permafrost and Periglacial Processes, 15(2), 171–188.

Washburn, A., 1979. Geocryology: A survey of periglacial processes and environments. Edward Arnold, London, 406 pp.

Washburn, A.L., 1956. Classification of patterned ground and review of suggested origins. Geological Society of America Bulletin, 67(7), 823–866.

Watanabe, T., Matsuoka, N., Christiansen, H.H., 2012. Mudboil and ice-wedge dynamics investigated by electrical resistivity tomography, ground temperatures and surface movements in Svalbard. Geografiska Annaler: Series A, Physical Geography, 94(4), 445–457.

Watanabe, T., Matsuoka, N., Christiansen, H.H., Cable, S., 2017. Soil physical and environmental conditions controlling patterned-ground variability at a continuous permafrost site, Svalbard. Permafrost and Periglacial Processes, 28(2), 433–445.

Westermann, S., Wollschläger, U., Boike, J., 2010. Monitoring of active layer dynamics at a permafrost site on Svalbard using multi-channel ground-penetrating radar. The Cryosphere, 4(4), 475–487.

Wilkinson, P.B., Meldrum, P.I., Chambers, J.E., Kuras, O., Ogilvy, R.D., 2006. Improved strategies for the automatic selection of optimized sets of electrical resistivity tomography measurement configurations. Geophysical Journal International, 167(3), 1119–1126.

Williams, P.J., 1961. Climatic Factors Controlling the Distribution of Certain Frozen Ground Phenomena. Geografiska Annaler, 43(3/4), 339–347.

Winsauer, W.O., Shearin Jr, H., Masson, P., Williams, M., 1952. Resistivity of brine-saturated sands in relation to pore geometry. AAPG bulletin, 36(2), 253–277.

Wolfe, S.A., Stevens, C.W., Gaanderse, A.J., Oldenborger, G.A., 2014. Lithalsa distribution, morphology and landscape associations in the Great Slave Lowland, Northwest Territories, Canada. Geomorphology, 204, 302–313.

Wright, N., Hayashi, M., Quinton, W.L., 2009. Spatial and temporal variations in active layer thawing and their implication on runoff generation in peat-covered permafrost terrain. Water Resources Research, 45(5), 13 pp.

Zenklusen Mutter, E., Phillips, M., 2012. Active layer characteristics at ten borehole sites in Alpine permafrost terrain, Switzerland. Permafrost and Periglacial Processes, 23(2), 138–151.

Zhang, T., Barry, R., Knowles, K., Heginbottom, J., Brown, J., 2008. Statistics and characteristics of permafrost and ground-ice distribution in the Northern Hemisphere. Polar Geography, 31(1–2), 47–68.

Zuidhoff, F.S., Kolstrup, E., 2000. Changes in palsa distribution in relation to climate change in Laivadalen, northern Sweden, especially 1960–1997. Permafrost and Periglacial Processes, 11(1), 55–69.

Zuidhoff, F.S., Kolstrup, E., 2005. Palsa development and associated vegetation in northern Sweden. Arctic, Antarctic, and Alpine Research, 37(1), 49–60.

Appendix A: Survey Details and Inversion Data Sets

Electrical Resistivity Imaging: q-3-D

Table A.1: List of Geophysical Surveys: q-3-D ERI.

Site/Label	Array Type	Number of Grid Lines	Electrode Spacing [m] (X/Y)	Number of Data Points for Inversion (of max.)	Survey Date
NAR/E01	D	17 (9/8)	2/3	6294 (6936)	12–14 AUG 2014
NAR/E24	WS+D	36 (20/16)	2/2	12441 (12472)	10–11 SEP 2016
UER/E01	WS+D	17 (9/8)	2/3	7372 (7800)	11–13 SEP 2014
UER/E02	WS+D	30 (17/13)	2/3	20523 (20880)	26–31 JUL 2015
UER/E03	WS+D	52 (14/12)	2/3	17813 (18096)	22–25 JUL 2016
LTF/E01	D	17 (8/9)	3/2	6111 (6936)	09–11 AUG 2014
FUR/E01	WS+D	22 (12/10)	2/3	10729 (11856)	30 SEP 2014–03 OCT 2014
HPA/E01	D	8 (8/0)	2/0	3264 (3264)	18–19 SEP 2015

D = Dipol-Dipol // WS = Wenner-Schlumberger

Electrical Resistivity Imaging: r-3-D

Table A.2: List of Geophysical Surveys: r-3-D ERI.

Site/Label	Array Type	Number of Electrodes (X/Y)	Electrode Spacing [m]	Number of Data Points for Inversion (of max.)	Survey Date
FUR/E01	WS+D	8/9	3	804 (816)	27 AUG 2015
FUR/E02	WS+D	8/9	3	731 (8162)	28 AUG 2015
OVR/E01	WS+D	8/9	1	816 (816)	17 SEP 2015
OVR/E02	WS+D	8/9	1	816 (816)	20 SEP 2015
OVR/E03	WS+D	8/9	1	816 (816)	16 SEP 2015
HPA/E03	D	8/9	2	497 (497)	19 SEP 2015

D = Dipol-Dipol // WS = Wenner-Schlumberger

Electrical Resistivity Imaging: ERTM

Table A.3: List of Geophysical Surveys: ERTM.

Site/Label	Array Type	Electrode Spacing [m]	Survey Length	Number of Data Points for Inversion (of max.)	Survey Dates
NAR/E02	D	2	70	2040 (2040)	12 AUG, 21 AUG, 29 AUG, 06 SEP, 06 OCT (all 2015)
NAR/E03	D	2	70	2856 (2856)	21 JUL, 28 JUL, 14 AUG, 25 AUG, 10 SEP, 29 SEP, 18 OCT (all 2016)
NAR/E08	D	1	35	2856 (2856)	21 JUL, 28 JUL, 14 AUG, 25 AUG, 10 SEP, 29 SEP, 18 OCT (all 2016)
NAR/E09	D	0.5	17.5	2856 (2856)	21 JUL, 28 JUL, 14 AUG, 25 AUG, 10 SEP, 29 SEP, 18 OCT (all 2016)

D = Dipol-Dipol

Electrical Resistivity Imaging: 2-D

Table A.4: List of Geophysical Surveys: 2-D ERI.

Site/Label	Array Type	Electrode Spacing [m]	Survey Length [m]	Number of Data Points for Inversion (of max.)	Survey Date
NAR/E04	WS	2	142	864 (864)	08 AUG 2017
NAR/E05	WS	3	105	288 (288)	09 AUG 2017
NAR/E06	WS	2	142	864 (864)	08 AUG 2017
NAR/E07	WS	2	142	1215 (1216)	10 AUG 2017
NAR/E10	WS	3	210	1216 (1216)	08 SEP 2017
NAR/E11	WS	3	105	288 (288)	12 AUG 2015
NAR/E12	WS	3	105	288 (288)	12 AUG 2015
NAR/E13	WS	3	105	288 (288)	12 AUG 2015
NAR/E14	WS	2	70	288 (288)	04 SEP 2015
NAR/E15	WS	2	70	288 (288)	10 AUG 2017
NAR/E16	WS	3	105	288 (288)	10 AUG 2017
NAR/E17	WS	3	105	288 (288)	10 AUG 2017
NAR/E18	WS	2	70	288 (288)	08 AUG 2017
NAR/E19	WS	2	70	288 (288)	07 AUG 2017
NAR/E20	WS	2	70	288 (288)	07 AUG 2017
NAR/E04	WS	2	142	864 (864)	08 AUG 2017
NAR/E21	WS	2	70	288 (288)	26 AUG 2016
NAR/E22	WS	2	70	288 (288)	25 AUG 2016

continues on next page

continued from previous page

NAR/E23	WS	2	70	287 (288)	14 AUG 2016
UER/E04	D	1	36	406 (408)	21 AUG 2015
OVR/E03	WS	2	142	1216 (1216)	17 SEP 2015
OVR/E04	WS	1	71	1216 (1216)	06 SEP 2017
OVR/E05	WS	2	70	288 (288)	20 SEP 2015
OVR/E06	WS+D	2	142	2926 (2926)	20 SEP 2015
OVR/E08	WS+D	2	70	696 (696)	16 SEP 2015
OVR/E09	WS	0.5	17.5	288 (288)	07 SEP 2017
OVR/E10	WS	0.5	17.5	276 (288)	07 SEP 2017
HPA/E02	D	2	70	408 (408)	19 SEP 2015

D = Dipol-Dipol // WS = Wenner-Schlumberger

Ground-Penetrating Radar

Table A.5: List of Geophysical Surveys: GPR.

Site/Label	Survey Type	Survey Length (2-D) or Extent (3-D) [m]	Frequency [MHz]	Survey Date
NAR/R01	1-D Common Mid-Point	—	100	20 JUL 2016
NAR/R02	2-D Common Offset	80	100	20 JUL 2016
NAR/R03	2-D Common Offset	45.25	100	20 JUL 2016
UER/R01	2-D Common Offset	300	50	23 JUL 2016
UER/R02	2-D Common Offset	100	50	23 JUL 2016
UER/R03	2-D Common Offset	80	50	23 JUL 2016
UER/R04	2-D Common Offset	74	50	22 JUL 2016
LTF/R01	1-D Common Mid-Point	—	50	09 SEP 2016
LTF/R02	1-D Common Mid-Point	—	50	09 SEP 2016
LTF/R03	2-D Common Offset	124	50	09 SEP 2016
LTF/R04	2-D Common Offset	52	50	09 SEP 2016
LTF/R05	2-D Common Offset	102	50	09 SEP 2016
LTF/R06	2-D Common Offset	102	50	09 SEP 2016
FUR/R01	2-D Common Offset	50	50	05 AUG 2017
FUR/R02	2-D Common Offset	43	50	05 AUG 2017
OVR/R01	3-D Common Offset	20x20	200	07 SEP 2017
OVR/R02	2-D Common Offset	72.5	100	06 SEP 2017
OVR/R03	2-D Common Offset	118	50	08 SEP 2017
OVR/R04	2-D Common Offset	20	200	06 SEP 2017
OVR/R05	2-D Common Offset	20	200	06 SEP 2017
OVR/R06	2-D Common Offset	45	200	06 SEP 2017
HPA/R01	2-D Common Offset	50	50	10 SEP 2017
HPA/R02	2-D Common Offset	50	50	10 SEP 2017
HPA/R03	2-D Common Offset	50	50	10 SEP 2017
HPA/R04	2-D Common Offset	28.1	200	10 SEP 2017

Temperature and Matric Potential Values

Table A.6: List of Recorded Time Spans Data Loggers.

Site/Label	Parameter	Recorded Time Span
NAR/T01	GST	17 Sep 2014–09 Aug 2017
NAR/T02	GST	17 Sep 2014–09 Aug 2017
NAR/T03	GST	17 Sep 2014–09 Aug 2017
NAR/T04	GST	17 Sep 2014–25 Apr 2016 and 14 Aug 2016–09 Aug 2017
NAR/T05	GST	17 Sep 2014–11 Aug 2016
NAR/T06	GST	17 Sep 2014–09 Aug 2017
NAR/T07	GST	15 Okt 2015–09 Aug 2017
NAR/T08	GST	17 Sep 2014–09 Aug 2017
NAR/T09	GST	17 Sep 2014–09 Aug 2017
NAR/T10	GST	17 Sep 2014–12 Aug 2016
NAR/Borehole	Subsurface Temperature	17 Sep 2014–10 Aug 2017
UER/T01	GST	26 Sep 2014–24 Aug 2017
UER/T02	GST	26 Sep 2014–24 Aug 2017
UER/T03	GST	26 Sep 2014–24 Aug 2017
UER/T04	GST	26 Sep 2014–24 Aug 2017
UER/T05	GST	26 Sep 2014–24 Aug 2017
UER/T06	GST	26 Sep 2014–24 Aug 2017
UER/T07	GST	26 Sep 2014–24 Aug 2017
UER/T08	GST	26 Sep 2014–24 Aug 2017
UER/T09	GST	26 Sep 2014–24 Aug 2017
UER/T10	GST	26 Sep 2014–24 Aug 2017
UER/Borehole	Subsurface Temperature	19 Sep 2014–24 Aug 2017
OVR/T01	GST	18 Sep 2015–23 Sep 2017
OVR/T02	GST	18 Sep 2015–23 Sep 2017
OVR/T02	Matric Potential	18 Sep 2015–23 Sep 2017
OVR/T03	GST	18 Sep 2015–23 Sep 2017
OVR/T03	Matric Potential	20 Sep 2015–24 Dez 2016
HPA	GST	19 Sep 2015–10 Sep 2017

GST = Ground Surface Temperature

Appendix B: Model Resolution and Survey Network Layouts

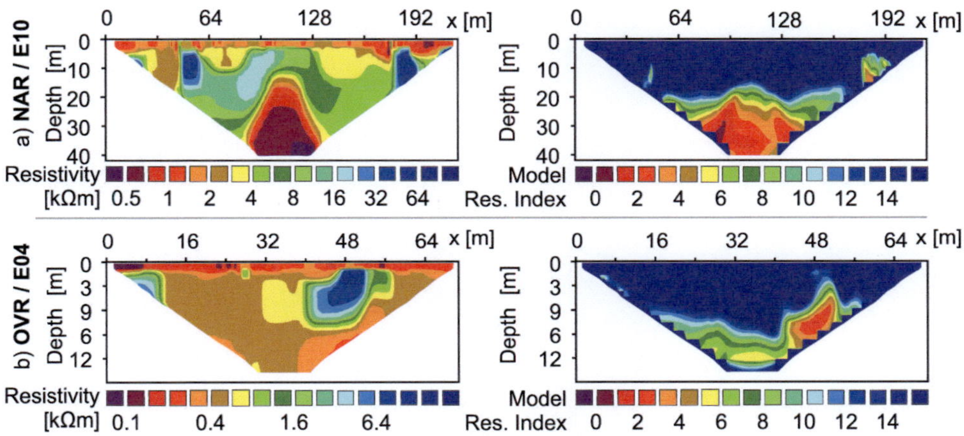

Figure B.1: Selected Inversion Models and Corresponding Model Resolution Index Values. Note that an Index Value of 10 is assumed to represent a lower limit for sufficiently resolved model cells (Loke, 2016a). a) Model NAR/E10, b) Model OVR/E04.

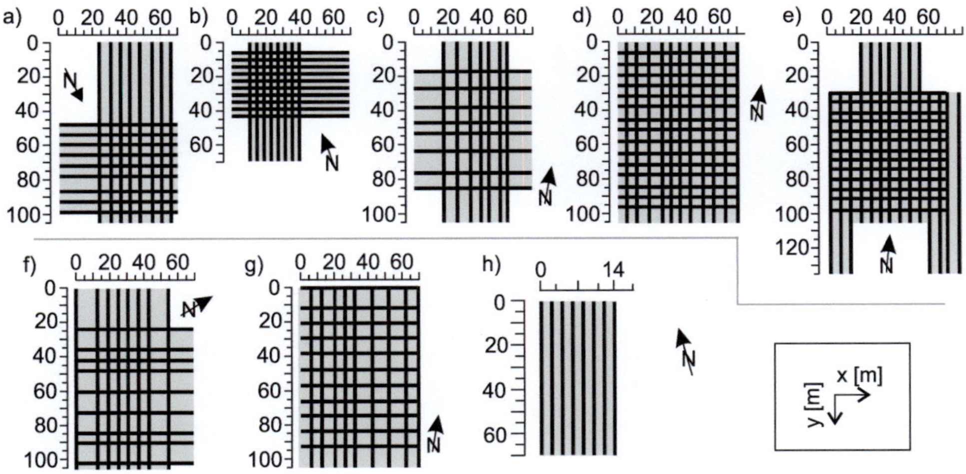

Figure B.2: Schematic Survey Network Layouts of q-3-D ERI Surveys. a) NAR/E01, b) NAR/E24, c) UER/E01, d) UER/E02, e) UER/E03, f) LTF/E01, g) FUR/E01, h) HPA/E01.